T0200012

Miroslava Ožvoldová
František Schauer

SPECTRUM SLOVAKIA Series
Volume 8

Remote Laboratories
in Research-based education
of real world phenomena

VEDA
PUBLISHING HOUSE
OF THE SLOVAK ACADEMY OF SCIENCES

PL | ACADEMIC
RESEARCH

Bibliographic Information published by the Deutsche Nationalbibliothek
The Deutsche Nationalbibliothek lists this publication in the Deutsche
Nationalbibliografie; detailed bibliographic data is available in the internet
at http://dnb.d-nb.de.

Authors: Miroslava Ožvoldová
                                      František Schauer

University of Trnava in Trnava
Faculty of Education
Slovak Republic
deprf@truni.sk

Monograph is a part of the activities within the projects of the Grant Agency
of the Slovak Republic Ministry of Education
KEGA No.011TTU-4/2012 "Energy as a category in science education via remote
experiments and integrated e-learning"
KEGA No. 020TTU-4/2013 "Accreditation of personalized ICT natural sciences
teacher's education centre"
APVV No.0096-11 "The role of defects in organic semiconductors for solar cells".

Reviewers: Prof. Dr. Javier Garcia Zubia,
                                      University of Deusto, Spain
                                      Dr. Gustavo R. Alves,
                                      Polytechnic of Porto, Portugal
Cover design and Layout   © Jana Sapáková, Layout JS.

ISSN 2195-1845
ISBN 978-3-631-66394-3   ISBN 978-80-224-1435-7
© Peter Lang GmbH   © VEDA, Publishing House
International Academic Publishers   of the Slovak Academy of Sciences
Frankfurt am Main 2015   Bratislava 2015

www.peterlang.com   www.veda.sav.sk

Dear reader,

The book you have in your hands deals with the two main scientific topics expressed in its title – remote experiments (RE) and remote experimentation, and utilisation of this new branch of information and communication technologies (ICT) in the strategy of education of research-based teaching and learning in elucidation of real world phenomena.

The book poses and tries to answer some basic questions connected with the subject matter of remote experimentation.

1. What are the properties of remote laboratories required for schools? Can Informatics cope with these?

What are the basic properties of REs required for schools? Decisive and important properties in the exploitation of REs at schools are reliability, fast communication via the Internet, simple and self-explanatory use, diagnostics and reports about the state of the RE, as well as reservation systems and data output for their evaluation. RE controlling programmes in JavaScript should enable reliable connection across the Internet, using any communicator at hand – smart telephones, book readers, iPads, notebooks and standard PCs.

To break down the barriers for RE preparation, the plug-and-play approach in the preparation phase and an expert system in the controlling programme as well as web page compiling without programming such as e.g. ER-ISES are needed.

2. Why the need to introduce remote experimentation and what are the recommendations of contemporary research?

The community of teachers and even the public ask – why is the teaching of natural sciences so ineffective and why does it brings very meagre results? In addition, the hours for natural sciences are constantly decreasing and the popularity of respective subjects at schools is diminishing. On the other hand, we have the results of extensive research on how to teach science disciplines effectively at our disposal, through shifting the learning focus from the institution to the student by activities like inquiry, group work, problem solving and connection to the real world and "learning by doing" using laboratories – in research-based courses.

In regard to the above-mentioned, REs may play a decisive role. Surprisingly, even though a great deal of research worldwide has been devoted to remote experimentation, especially remote e-laboratories and despite the substantial progress achieved, the proportion of REs in teaching is very low.

3. How can we use remote experimentation in research-based education?

The obstacles in using REs in teaching are manifold. The present discussion about new teaching methods is no longer directed towards the fundamental changes in the learning processes induced by the new ICT, but towards the method of introducing the new techniques into everyday teaching practice by establishing the resources of e-learning, curricula, etc. We also adhere to the opinion that laboratories and simulations can radically change education in physics via new strategies and new teaching tools. We therefore suggest the method of e-learning including remote experimentation, i.e. Integrated e-Learning (INTe-L).

4. How can we help teachers in using remote experimentation?

This may be achieved by building the Remote Laboratory Management System (RLMS) and providing ample services for teachers as a database of RE with the corresponding tools and accompanying teaching materials,

Illustration photo _ Students of the Faculty of Applied Informatics TBU in Zlin. In the research-based physics laboratory with hands on, virtual and remote experiments using INTe-L strategy

their reservation system, diagnostics of RE, inserting simulations, using IP telephony and a whiteboard, and the set-up of virtual classes.

The introduction of remote laboratories and their full use in research-based teaching as well as the preparation of future teachers of science disciplines at Faculties of Education is of the utmost importance. In addition, in-service teachers should attend courses to learn how to use ICT in teaching in general and RE in particular, and building their own RE using new environments for doing so.

The book is structured according to the questions posed, trying to find at least partial answers to them. Chapter 1 summarises the reasons for the changes towards research-based education. In Chapter 2, we deal with the technical aspects of REs built on the Internet School Experimental System (ISES) hardware, and the possibilities of controlling programmes and web pages compiled by means of Easy Remote-ISES (ER-ISES expert environment). In Chapter 3, the system of incorporating RE into research-based teaching by means of a new strategy of education – Integrated e-Learning (INTe-L) is presented. In Chapter 4 we summarise examples of research-based physics teaching, using not only remote experiments, but also an interactive blackboard, clickers and LMS MOODLE. In Chapter 5 we deal with the Remote Laboratories Management System (RLMS) that serves to manage the great number of remote laboratories from many working places all across the European Union. RLMS should provide RE to the teacher in a straightforward way by organising booking and supervising their functions such as diagnostics, delivering accompanying material and tools for RE and instantaneously forwarding information needed for teaching.

The book is devoted to all those interested in using ICT as a means of education in general and remote experimentation in particular, as well as to/for those interested in student-centred activities by ICT.

We have pleasure in thanking both the co-operating workplaces of the Consortium, namely the Faculty of Mathematics and Physics of the Charles University in Prague and the Faculty of Applied Informatics of Tomas Bata University in Zlin, whose activities were indispensable in presenting the results attained and shown here in the current book. Special thanks also go to Associate Professor Dr. František Lustig for his vision in the branch of computer-oriented laboratories, and for his many years of fruitful co-operation in the field of using ISES in remote experimentation. We also wish to thank Mr. Jiři Dvořák for his enormous programming work on the ISES remote experiments software.

František Schauer, Miroslava Ožvoldová

List of Abbreviations and Acronyms

ABET	Accreditation Board for Engineering and Technology
ADDA	Analogue-Digital, Digital-Analogue Transducer
COLLES	Constructivist On-Line Learning Environment Survey
CHU	Charles University in Prague, CZ
DW	Data warehouse
EJS	Easy Java Simulations
ER ISES	Easy Remote ISES
FCI	Force Concepts Inventory
HW	Hardware
ICT	Information and Communication Technologies
INTe-L	Integrated e-Learning
IPS	Intrusion Prevention System
ISES	Internet School Experimental System
ISES WIN	Software for laboratory hands-on ISES experiments
JSON	JavaScript Object Notation
LMS	Learning Management System
MOODLE	Modular Object Oriented Dynamic Learning Environment
PHP	Hyper-Text Pre-processor
RE	Remote Experiment
RL	Remote Laboratories
RLMS – REMLABNET	Remote Laboratory Management System built by Consortium of CHU in Prague, TBU in Zlin and TU in Trnava
RTSP	Real Time Streaming Protocol
SIEM	Security Information and Event Management
SW	Software
TBU	Tomas Bata University in Zlin, CZ
TEDS	Transducer Electronic Data Sheets
TU	Trnava University in Trnava, SK
TCP/IP	Transmission Control Protocol
VOIP	Voice over Internet Protocol
XML	Extensible Mark-up Language

Contents

Remote laboratories worldwide and in Slovakia

It was at CERN that Tim Berners-Lee developed and installed the first web-server in 1991. By that year, Aburdene et al. envisaged "Laboratory experiments being operated remotely and shared among universities "[1]. Although at that time there were discussions in the USA, when CERN decided to freely open the WWW to anyone who had the opportunity to share knowledge and start working together, without restrictions, in a collaborative way through the Internet.

"The Internet has revolutionised the computer and communications world like nothing before. The invention of the telegraph, telephone, radio, and computer set the stage for this unprecedented integration of capabilities. The Internet immediately offers worldwide broadcasting capability, a mechanism for information dissemination, and a medium for collaboration and interaction between individuals and their computers without regard for geographic location"[2].

The first remote experiment for teaching purposes was announced in the mid 1990's [3]. The following years witnessed considerable growth in the activities related to using the web for supporting education in general, and also experimentation in particular. The list is very exhaustive and may be found in [4].

The present state of the development of Information and Communication Technologies (ICT) is bringing about major qualitative changes. In the editorial in the issue of Eur. J. Phys. devoted to student undergraduate laboratory and project work, D. Schumacher brings examples of the invasion of computers in contemporary laboratory work, including project labs, modelling tools, interactive screen experiments, remotely controlled labs, etc. Schumacher closes with the plausible statement: *"One can well imagine that project labs will be the typical learning environment for physics students in the future."* [5]. The same goes for remote experiments (RE) which have spread across the Internet and contemporary literature is extensive (e.g. recent monographs and references [6][7]).

Surprisingly, the use of RE for education purposes and for special groups of students such as students of open universities, distance students and those who, for various reasons, cannot attend traditional school has, until recently, scarcely been dealt with. Tkáč and Schauer [8] discuss the advantages of using RE for distance students, for whom laboratory work in the first terms of physics in technology-oriented universities is rarely available, and physics education means lectures and seminars only. The authors suggest the introduction of remote experimentation delivered from an e-laboratory as one of the possible solutions.

In the light of emerging simulated and remote engineering laboratory courses, the U.S. Accreditation Board for Engineering and Technology (now renamed ABET) has taken on the task of assessing whether these new courses can truly accomplish the goals of educational laboratories and are an adequate substitution for traditional hands-on experience. A list of 13 learning objectives designed for this purpose describes the goals of an engineering laboratory [9]. In our case, the general and the decisive criterion for the introduction of e-laboratories was the trend to involve students at a deep level in practical experimental work, and to remove the barriers for independent laboratory work. The provision of remotely controlled real experiments accessible over the Internet can potentially and positively address all these issues, since:

- Remote real experiments in most cases work in a round-the-clock regime. Students and those interested may choose the optimum time for them and work at a pace most suitable for them. If an instructor's presence is required, then the reservation system for experiments is needed.
- They enable access to costly and potentially risky experiments.
- They introduce students to the real world of science and technology, with teamwork mediated by computers and the web, and organised at a distance, thus building correspondence skills.
- The actual university premises do not limit access to RE, since the experiments are mediated by the computer and the Internet. A group of distance students may profit from RE. In the present scheme of organisation, they work in laboratories in time blocks covering more than one laboratory.
- RE positively influences all groups of people who, for various reasons, are otherwise excluded from laboratory work.
- The remote experiments support the idea of globalisation and delivery of experiments to all destinations without exception.

The first two remote experiments in Slovakia occurred nearly simultaneously:

1) In the field of automatic control in the Department of Automatic Control, Faculty of Electrical Engineering of the Slovak University of Technology in Bratislava, developed by the team of Prof. M. Huba in 2007 (see K. Žáková [10] and P. Bisták [11]),

2) In the field of "Electrochemical cell" in 2008, in the workplace of the Department of Physics, Faculty of Education in the University of Trnava in Trnava, team of Prof. F. Schauer and Assoc. Prof. M. Ožvoldová (see L. Válková and F. Schauer [12] and L. Válková, F. Schauer and M.

Ožvoldová [13]). Placed on the net since then, the experiment is being constantly improved and modernized, http://remotelab2.truni.sk.

In addition, some remote experiments in Slovakia have been built by certain groups as a result of common activities in the commom solved projects oriented through the remote laboratories: the group of Prof. P. Ballo of the Department of Physics, Faculty of Electrical Engineering of the Slovak University of Technology in Bratislava in the field of "Characteristics of thermal emission", Dr. K. Kvetan on "Electronic circuits" from the Department of Physics, Faculty of Materials Science and Technology of Slovak University of Technology in Bratislava, and Prof. T. Kozík with the experiment on hydrodynamics of the Department of Technology and Information Technologies, Faculty of Education of Constantine the Philosopher University in Nitra. Very recently (2013), some remote experiments were bought by the Department of Physics, Faculty of Natural Sciences of the same University.

The topic of remote experiments has been the core of scientific activities of the Department of Physics, Faculty of Education in the Trnava University in Trnava since 2005, when the first remote experiment was built. We relied on the enormous know-how of Assoc. Prof. F. Lustig from the Department of Physics Education, Faculty of Mathematics and Physics, Charles University in Prague, where the universal and very useful modular computer-oriented set of the Internet School Experimental System (ISES) was designed at the beginning of the 1990's [14]. Up until today, several generations of ISES have been built, such as a recent USB ISES for use with notebooks and one chip USB ISES devices.

In 2009, we established a Consortium of three universities: the Faculty of Mathematics and Physics, Charles University in Prague, the Faculty of Applied Informatics, Tomas Bata University in Zlin and the Faculty of Education in Trnava University in Trnava with the aim to build common remote laboratories and a laboratory management system within the European Union.

Since then, , about 20 remote experiments have been built and utilized in Trnava. They are available at the following URL addresses http://remotelabN.truni.sk, N = 1, 2, ..., 20). Their functionality and technical and didactical use will be described in Chapter 3 and in the Appendix. Figure P1 shows the tabular leaflet with the contemporary Trnava REs, Figure P2 shows the leaflet with the Prague REs, illustrating the situation in the year 2013, when all the experiments were running under Java applets. Since then, when the company Oracle prohibited forwarding the remote experiments by using the latest versions of Java software, we had to resort to another web

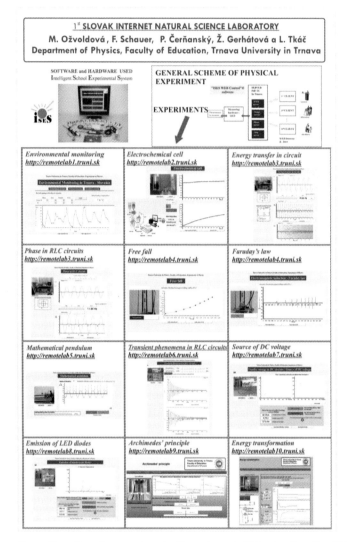

Figure P1 _ Remote experiments, Department of Physics, Faculty of Education, Trnava, (June 2013) url addresses are correct, the web pages were changed in november 2014

communication language. After careful analysis, we chose JavaScript as a reasonable option [15]. Fortunately enough, we have changed the already existing ER-ISES environment also to JavaScript language, so that, since August 2014 all the remote experiments of the Trnava University have been

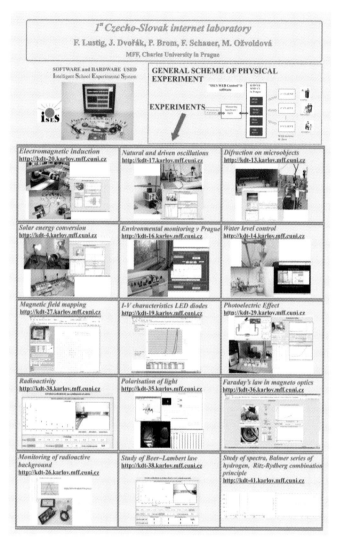

Figure P2 _ Remote experiments, Faculty of Matthematics and Physics, Charles University in Prague (the situation in 2013)

reprogrammed using ER-ISES to JavaScript and those will be presented in this monograph. The same progress has taken place at the Charles' University and the process of transferring the remote experiments there to JavaScript has been finished.

Chapter 1

Research-based education and experiments

Remote laboratories (RL) and remote experiments (RE) are the monograph's leading topic. Before we embark on the main topic, let us explain the reason why we consider RE and concomitant necessary changes in teaching as an imperative in the contemporary education system. Our reasoning stems mainly from our extensive experience in teaching physics and persuasion about the necessity of changes in traditionally based education, where experimentation and remote experimentation along with other means of information and communication technologies may exert profound changes towards a science-oriented approach.

1.1 Traditional courses versus research-based courses

Traditional courses of sciences

The main qualities of the prevailing traditional higher education can be summarised into five points [16]:

1. The education goals of modern society are quite different from what they used to be – complex problem solving that is far more difficult than simple information knowledge and the prevailing system is not economically practical for large scale use.
2. We are facing an unprecedented educational challenge: the need to teach complex technical knowledge and skills effectively to a large section of the total population.
3. The responsibilities of the faculty teaching staff are far more different than they used to be several decades ago. A modern research university nowadays plays a major role in knowledge acquisition and application in science and engineering. Running a research program has become an inevitable and most recognised activity of nearly every member of the science and engineering faculty.
4. "Research has established that people do not develop a true understanding of such a complex subject like science by listening passively to explanations" [17]. True understanding only comes if students actively construct their own understanding in the process of mentally building upon their prior thinking and knowledge [18]. This construction of learning is dependent on the epistemologies and beliefs they bring to the subject and these are readily affected (positively or negatively) by instructional practices and the student's beliefs [19].

5. The final dramatic change is in the state of education-related technologies [20] (see Figure 1.1).

As a community, we must ask: *"How successfully are we educating students in the Sciences?"*[21]. Data indicates that we are not where we want to be, as too many undergraduates in our courses are not learning the science [22]. We are fortunate, however, that we have access to a growing body of research on effective ways of teaching science [17][20][23][24][25]. Research tells us how to improve student learning through student-centred activities such as inquiry, peer instruction, and group work, while focusing on problem-solving and concepts related to the real world. However, most undergraduate science courses are taught via lectures [26]. How can a teacher use interactive techniques to restructure an existing course? [27]

In principle, we may observe four learning puzzles in teaching [28].

Puzzle 1 – Retaining information

As an example [29], C. Wieman frequently presented a non-obvious fact in a lecture along with its illustration, and then quizzed the students 15 minutes later on the fact. About 10 percent usually remembered it by then. To see whether they simply had mentally deficient students, C. Wieman repeated

Figure 1.1 _ Illustrative photos of the use of information communication technologies at three levels of education

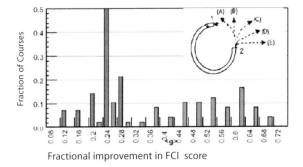

Fractional improvement in FCI score

Figure 1.2 _ Comparison of understanding the basic concepts – traditional courses (red) and research-based courses (green) [32], with kind permission of AJP

that experiment when giving a departmental colloquium at one of the leading physics departments. The audience was made up of the physics faculty members and graduate students, yet the result was about the same – only around 10 percent retained the information. It is a result of cognitive science: an extremely limited capacity of the short-term working memory. Research tells us that the human brain can hold the maximum of about seven different items in its short-term working memory and can process no more than about four ideas at once.

Puzzle 2 – Understanding Basic Concepts

The second puzzle rests upon another observation [30]. The idea was to study how well students are actually learning the basic concepts in their physics courses, using Force Concepts Inventory (FCI) [31]. This instrument tests students' mastery of the basic concepts of force and motion, which are covered in every first-semester postsecondary physics course.

Richard Hake compiled the FCI results from six thousand participants of traditional courses [32] (Figure 1.2) and found that, in the traditional lecture course, students master no more than 30 percent of the key concepts that they did not know at the start of the course (shown in red). Similar results are reported in many other studies, largely independent of lecturer quality, class size, and institution. Pedagogical approaches involving more interactive engagement of students show consistently higher gains on the FCI and similar tests (shown in green).

Puzzle 3 – Affecting beliefs about science [33]

The novice/expert-like beliefs about science in general and his/her field of studies are important in many ways, e.g. content learning and interest in the field of study. It is important to stress why this topic is worth studying, how it works in the real world and how it is connected to the things the student has been acquainted with. This approach influences enormously the attitudes of students and their study results.

Puzzle 4 – Teaching and learning improved by using technology

The teacher has to master classes of both 10 and 200 students. Use of clickers [34], simulations [35] and remote experiments helps enormously in both situations. Using these various effective pedagogical strategies, we may see dramatic improvements in learning.

The above-mentioned puzzles can be enhanced by another one: A graduate student enrols in postgraduate study. Why did his/her first 17 years of education seem so ineffective, while a few years of doing research turned a graduate student into an expert physicist? The first part of the mystery is solved: the traditional science courses did little to develop expert-like thinking about physics. But what makes working in a research lab so different? Let us answer this question as well.

Research-based courses, or how to teach better [29][36]

The current approach to education is built on the assumption that students come to school with different brains and that education is the process of immersing those brains into knowledge, facts, and procedures, which those brains then absorb to varying degrees. The extent of absorption is largely determined by the inherent talent and interest of the brain. Thus, those with "talent" will succeed, usually easily, whereas the others have no hope. Recent research into these questions discovered that talent is overrated [37]. Research advances in cognitive psychology, brain physiology, and classroom practices are painting a very different picture of how learning works. We are learning that complex expertise is not a matter of filling up an existing brain with knowledge, but of brain development. This development comes about as the result of intensive practice of the cognitive processes that define the specific expertise, and effective teaching can greatly reduce the impact of initial differences among the learners. This research has established important underlying causes and principles and important specific results, but it is far from complete. More research is needed on how to accomplish the desired learning most effectively over the full range of skills and potential learners in our classrooms, as well as how to best train teachers.

A lot of educational and cognitive research can be reduced to the following basic principle: People learn by creating their own understanding by deliberate practice. However, that does not mean they must, or even can do it without assistance. Effective teaching facilitates that by getting students engaged in thinking deeply about the subject at an appropriate level and then monitoring that thinking and guiding it to be more expert-like.

As an answer to the four above-mentioned puzzles, let us follow the trajectory of improved teaching.

Reducing Cognitive Load

Cognitive science gives straightforward proofs about the extremely limited capacity of the short-term working memory. Research tells us that the human brain can hold a maximum of about seven different items in its short-term working memory and can process no more than about four ideas at once [17]. Therefore, anything the teacher can do to reduce that cognitive load while presenting the material will help. Some ways to do so are obvious, such as slowing down. Others include having a clear, logical, explicit explanation to the class (including making connections between different ideas presented and connections to things the students already know), using figures where appropriate rather than relying only on verbal descriptions and minimising the use of technical jargon.

Stimulating and Guiding Thinking

The learning of complex expertise is therefore quite analogous to muscle development. In a similar way, the brain changes and develops in response to its strenuous and extended use. First, it means that learning is inherently difficult, so that motivation plays a large role. To succeed, the learner must be convinced of the value of the goal and believe that hard work, not innate talent, is critical. Second, activities that do not demand substantial focus and effort provide little educational value. Listening passively to a lecture, doing many easy, repetitive tasks, or practicing irrelevant skills produce little learning. Third, although there are distinct differences among learners, for the great majority the amount of time spent in deliberate practice transcends any other variables in determining learning outcomes. An example of how teaching and learning can be improved is by implementing the principle that effective teaching consists of engaging students, monitoring their thinking, and providing feedback. Given the reality that student-faculty interaction at most colleges and universities is going to be dominated by time spent together in the classroom, this means the teacher must make this happen first and foremost in the classroom. Getting students engaged and guiding their thinking in the classroom is just the beginning of true learning, however. This classroom experience has to be followed up with extended "effortful study," where the student spends considerably more time than is possible in the classroom developing expert-like thinking and skills. To ensure that the necessary extended effort is made and that it is productive requires carefully designed homework assignments, grading policies, and feedback.

Addressing Beliefs

The particular intervention that addresses student beliefs is by explicitly discussing for each topic covered why this topic is worth learning, how it operates in the real world, why it makes sense and how it connects to things the student already knows.

Using Technology

In this respect, the final dramatic change is in the state of education-related technologies. Everyone is aware of the enormous increases in the capabilities of information technology (IT) over the past few decades, years, and even months. These offer many fairly obvious opportunities for dramatically changing the way teaching is carried out in colleges and universities, and in the process, making higher education far more effective and more efficient. Unfortunately, these vast opportunities remain largely untapped. While there are a few spectacular examples, generally the educational IT currently available is quite limited in both quantity and quality [6].

Using clickers – each student has a clicker to answer questions posed in the class. A computer records each student's answer and displays a histogram of those responses. The clicker quickly and efficiently gets an answer from each student for which that student is accountable but which is anonymous to their peers [34][38]. The most productive use of clickers in our experience is to enhance the Peer Instruction technique developed by Eric Mazur, particularly for less active or less assertive students [39].

Another powerful educational technology is the sophisticated online interactive simulations [40 – 43]. This technique can be highly effective and takes less time to incorporate into instruction than more traditional materials. The Colorado group [40] has created and tested over 60 such simulations and made them available for free. In our teaching, we have explored their use in the lecture and for homework and as replacements for, or enhancements of, laboratories and examinations.

The third group of ICT conditioned tools for teaching are remote experiments in Remote Laboratories (RL). (For summarized information on remote laboratories and their use see recent monographs [6][7][44].) The subject is of growing interest allowing a single common pool of innovative science apparatus to be shared across many schools and thus many students [45][46]. Consider, for example, an item of laboratory apparatus that is used for two 1-hour demonstration sessions each term. If this apparatus was made remotely accessible, and supported by an appropriate scheduling system, then instead of supporting a single school, it could support up to 125 schools (10 weeks x 5 days x 5 hours per day/2 hours usage per school), and possibly many more if the schools existed across multiple time zones. The overall conclusions from the research indicated that remote laboratories, if used appropriately in a way that takes into account the intended educational outcomes of the laboratory experience, can provide significant economic benefits.

However, beyond the clear benefit of improved utilisation, there is a range of other benefits, including access to a wider range of equipment than would be feasible for individual schools to support, enhanced accessibility for students in terms of both their time and location of access and improved opportunities and efficiencies for gathering and analysing rich experimental data. Although a significant body of literature currently describes specific examples of the use of RL, the focus has been dominated by the higher education sector only, potentially because those developing the laboratories have generally worked in that sector. Much less consideration has been given, at least in the published literature, to the use of RL in primary and secondary schools [47].

1.2 Photo gallery of our remote experiment activities

Figure 1.3 _ Students of Faculty of Applied Informatics TBU in Zlin constructing the remote experiment "Water level control" (photo F. Schauer, 2010)

Figure 1.4 _ Students of Faculty of Applied Informatics TBU in Zlin constructing
the remote experiment "Solar energy conversion" (photo F. Schauer, 2011)

Chapter 2

ISES remote laboratories

2.1 ISES remote experiments

The conventional and traditional approach to the teaching of scientific subjects and information dissemination, oriented at both secondary school and university students, or a broad public, are quite outmoded and not at all popular. Contemporary students, as digital natives, demand a higher level of teaching methods which help to perceive phenomena of the real world, delivered by natural and technology oriented sciences. There is also the factor of accessibility of tools and means for all sorts of students, who frequently prefer to study scientific subjects via the Internet, available as commonplace on their computers. With this development several important new ICT tools emerged. One of them is RE, formed by globally accessible e-laboratories across all time zones 24/7, and in the majority of cases free of charge (the present state of remote laboratories may be found in recent monographs [6][7]). But unfortunately, their physical SW and HW, as well as the informatics SW and HW differ widely [48][49].

The RL of the Consortium (comprised of three partners, Trnava University in Trnava, Tomas Bata University in Zlin and Charles University in Prague) are built on ISES [50] that has been entirely developed for educational purposes. The ISES is a complex tool for real time and remote data acquisition, data processing and control of experiments and other processes. It is an open system consisting of a basic ISES HW with the ISES WIN SW intended for a local computer oriented laboratory. The system, from its very start (the beginning of the 1990's), was designed for controlling experiments and later even for RE. The ISES WEB Control Kit (built in 2000) served this purpose for remote counterpart building. Recently, the environment for the intuitive compiling of control programs for remote experiments was written. ER-ISES is like a superstructure based on ISES equipment and developed in cooperation with TBU in Zlin and Charles University in Prague (in 2012) [51].

Let us describe the functioning principle of the plug-and-play ISES in the building of hands-on experiments and their transformation to remote.

Internet School Experimental System (ISES)

ISES is a software and hardware set for plug-and-play school experimentation with the general scheme depicted in Figure 2.1. The ISES system was described in detail elsewhere [52] and here we give only a few relevant details. The physical HW is composed of the ISES panel and the set of

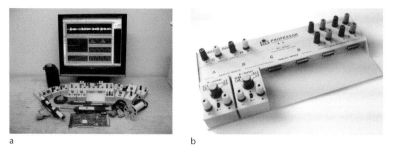

a   b

Figure 2.1 _ Plug-and-play physical HW of Internet School Experimental System system (ISES): the adda card and the set of ISES modules (a), the ISES panel with two ISES modules (b) (see also http://www.ises.info)

modules and sensing ISES elements (Figure 2.1 a) easily interchangeable, their presence and adjusted range are automatically sensed by the computer, with the automatic calibration facility. The ISES panel (Figure 2.1 b) enables 10 different channels (6 analogue and 2 digital) and a capability of 2 programmable outputs. The informatics HW is composed of the interface card with A/D and D/A converters with about 40 ISES modules and sensors and as the informatics SW the controlling programme ISESWIN. On the right side the evaluation panel is shown with the possibility of time and frequency measurements, curve fitting, integration, differentiation, etc.

2.2 ISES remote experiments architecture

Hands-on ISES experiments may be easily transformed to the RE. The ISES hands-on experiment with physical hardware (HW) and informatics software (SW), (Figure 2.2 a), remains identical, only the informatics SW is

Figure 2.2 _ ISES hands-on to ISES remote experiment transformation: hands-on experiment (a), remote experiment (b)

a   b

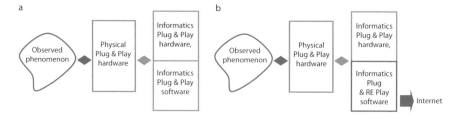

changed to build ISES RE (Figure 2.2 b). The controlling system for the ISES RE operation is depicted in Figure 2.3 [62].

The first component of the RE is the apparatus or physical HW, which is the most important hardware component, including all the equipment that examines and measures the phenomenon in question. This physical HW contains ISES probes, sensors and meters. For example, ISES volt-meter, ISES ampere-meter, ISES ohm-meter, ISES flow-meter, etc. Physical HW is wired to the ISES panel which transfers incoming electrical signals to the PCI A/D-D/A card installed on the computer to gather and process the data.

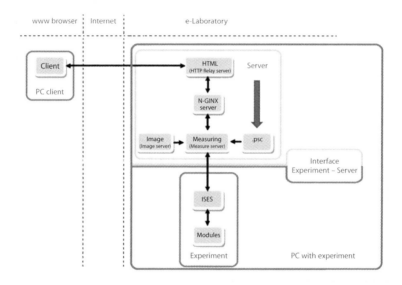

Figure 2.3 _ Finite-state machine controlling system for the ISES re operation

The second component is the Measureserver module, the most important informatics SW component of the remote experiment, functioning as a finite-state machine (FSM) [53]. It communicates with the apparatus, processes the measured data and controls commands. The main feature of the Measureserver is the setting of the ISES panel, the sensors/meters for data collection and processing the control commands. The module dynamically responds to signals from the physical HW, as well as the commands transmitted from the client's interface. Measureserver has three main components – the Measureserver core, Hardware plug-in and PSC script file. The

Measureserver core is responsible for the data and command transfer, client serving and for the execution of all the controlling commands, which is controlled by the PSC script file, which is directly imported to the Measureserver. The PSC script is a unique programming language specially designed for the ISES system. The PSC script is a non-compliable language defining the core behavior of the Measureserver remote experiments. The hardware plug-in provides required functionality to control the ISES panel translating signals from/to the physical HW apparatus [51].

The third component is the ImageServer. This component serves for the live view of the remote experiment. This module periodically saves the actual view of the realized experiment from the image pick-up device, usually a USB web camera, to the storing file. The filename, resolution and quality of the picture are easily adjusted during the installation process [51]. The fourth component is a web server for the communication between RE and a client. At present, Nginx is used, due to sufficient stability after loading an appropriate webpage of the remote experiment on the client's side computer to start the experimentation. The fifth and last component is the RE communication webpage, which is the interface communicating with the RE via the Internet. Such a webpage can be displayed by all standard web browsers on computers where various operating systems can be running.

Figure 2.4 _ Schematic arrangement of client-server communication of ISES remote experiment

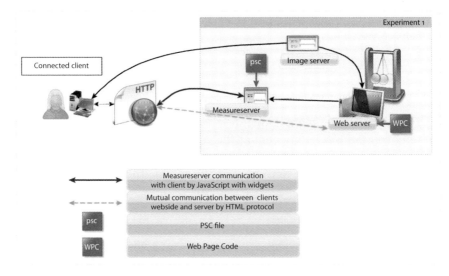

Based on the system of RE, clients have non-stop accessibility to enter RL through their web browser's interface connected to the Internet from anywhere (Figure 2.4). The most significant advantage is the real-time experimenting with the apparatus installed in a laboratory. An authorised student can comfortably communicate with the remote apparatus of the visualised RE via a webpage. The apparatus promptly reacts and sends adequate responses/signals/data through particular subsystems back to the target client/student. After the completion of RE the student will arrange the data in formatted, sorted and filtered form and also in graphic charts displayed on the screen and find the corresponding answers regarding the phenomena observed.

2.3 Example of ISES plug-and-play remote experiments

The EU FP7 work programme for 2012 in the call ICT-2011.8.1"Technology-enhanced learning supporting European-Wide Federation and use of remote laboratories and virtual experimentation for learning and teaching purposes" is an objective devoted to remote laboratories and assigns, as one of the main tasks, to deliver and carry out research on "Open interfacing components for easy plug-and-play of remote and virtual labs should be made available to stimulate the growth of the network of labs." [54] The significance of this reasoning for the plug-and-play systems is obvious. It is governed by the leading idea of disseminating the e-laboratories among the wider population, not keeping them for the small group of "devoted".

On the other hand, to some researchers working in the field of PC oriented experimentation and remote experiments (RE), the connection of RE plug-and-play interfaces may seem as a mere construction, unexpected and without any sense. The RE is by now a rigid system, which uses fixed physics hardware and has to create the controlling programme for the informatics HW and SW, based on the flow diagram, enabling only the programmed development of the experiment and instantaneous controls settings of the RE.

To bridge the gap and contribute to the development of remote experimentation, we try to show the possibility of plug-and-play, especially in the physical hardware and sensors and measuring instrumentation and show the initial results achieved in designing the plug-and-play system ISES with the example of the ISES RE plug-and-play of the mechanical harmonic oscillator. Let us explain the plug-and-play concept first.

Plug-and-play in experimentation

"Plug-and-play is used to describe devices that work as soon as they are connected to a computer system. The user does not have to manually install drivers for the device or even tell the computer that a new device has been added. Instead, the computer automatically recognizes the device, loads new drivers for the hardware if needed, and begins to work with the newly connected device." [55.] This definition of plug-and-play comes from the PC and was enforced by the fact that the process of configuring devices, forming the PC or peripheries manually could be quite difficult, and there was usually no forgiveness for technical inexperience. Incorrect settings could render the whole system or just the expansion devices completely or partially inoperable.

The process of measurements may be divided into the sensor recognition and calibration and measuring processes itself. Here the plug-and-play approach is more or less applied with every measuring instrument. In plug-and-play measurements the situation is not so obvious and producers of industrial applications concentrated on the design of interfacing the measuring instruments to the PC using plug-and-play interfaces (IEEE-484, USB, PCI, Express Card and even wireless IEEE 802.11) for connecting plug-and-play smart sensors. A typical example is NI approach (http://www.ni.com/teds) using virtual Transducer Electronic Data Sheets (TEDS) technology where traditional analogue sensors may be configured electronically using data files for identification, configuration, and calibration information storing for an individual sensor.

Plug-and-play in a hands-on school experiment

Computer oriented experimentation school systems (ISES, Vernier, Pasco, IP Coach, Lego, and also LabVIEW) have been designed for traditional laboratory school measurements. The systems are designed on the unified basis of the A/D-D/A card as the measuring interface, or alternatively USB, Bluetooth, Zig Bee, WiFi interfaces. The measurement systems ISES, Vernier, Pasco, IP Coach and others have their sets of sensor modules with plug-and-play connection and auto detection and range detection, some of them with automatic range adjustment, extraordinarily equipped with automatic modification of software (e.g. for heart rate the frequency display or for signal the Fourier transformation, etc.). LabVIEW is a specific measurement system, which is more or less software oriented, communicating with many measurement and control cards and interfaces. It is not equipped with the special set of measurement sensor modules, on the contrary, it can, after a small amendment, communicate with almost all enumerated measurement school systems.

Plug-and-play in a ISES remote experiment

Today, when remote experiments spread across the Internet and start to become a very important segment of ICT in education, there are only two systems with the SW support for creating RE – LabVIEW and ISES. The former has the disadvantage of the necessity to programme the plug-and-play modules and install LabVIEW Run-Time Engine of the latest version to run the RE (http://joule.ni.com/nidu/cds/view/p/id/2087/lang/de). This may be prohibitive in school networks and for beginners in the field.

Let us discuss the possibility of creating RE by ISES in more detail. In Figure 2.5 the ISES hands-on experiment of the mechanical oscillator with the physical plug-and-play hardware measuring system and data output in graphical and tabular forms is shown. The set-up of the hands-on experiment of the mechanical oscillator with mass and spring, where we use the ISES dynamometer for the measurement of the force acting on the suspension (and thus also the instantaneous deflection), is depicted in Figure 2.5. The time dependences of the oscillations on the ISES standard display in two channels with two sensitivities are visible.

Once we have the hands-on experiment of mechanical oscillations, it is straightforward to transform it into the RE version using the ISES WEB CONTROL kit. This is the software package of plug-and-play units that contains a library of 25 common situations for the set-up of the controlling

Figure 2.5 _ ISES hands-on experiment of mechanic oscillator, the setup with mass, spring, ISES dynamometer (a) and the data outputs (b)

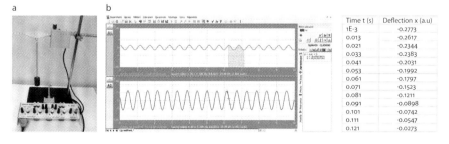

Time t (s)	Deflection x (a.u)
1E-3	-0.2773
0.013	-0.2617
0.021	-0.2344
0.033	-0.2383
0.041	-0.2031
0.053	-0.1992
0.061	-0.1797
0.071	-0.1523
0.081	-0.1211
0.091	-0.0898
0.101	-0.0742
0.111	-0.0547
0.121	-0.0273

RE webpage as time dependence of a quantity, mutual dependence of two quantities, the ruler, the data displays etc., all with or without a web camera. As a matter of fact, these are the pre-programmed JavaScript widgets, which have several input parameters to adjust their function.

In Figure 2.6 a small part of the ISES WEB CONTROL library (in Czech language until now) is shown on the left. We present also its application on two web pages of typical RE: the data transfer of the time dependence of the temperature (shown above right) and the other on the movement in two directions, measured by the physical hardware ISES (shown below right).

By using CD with the ISES WEB CONTROL kit containing the library, it is easy to transform any simple experiment to its remote version by the copy and paste principle. Our mechanical oscillator hands-on experiment was transformed by a similar procedure, as is shown in Figure 2.6. There are two RE versions of the experiment – in Figure 2.7 the RE with graphical time dependence of the oscillations with the camera view and in Figure 2.8 the RE with graph and data transfer to the researcher PC, both with data output is represented.

Figure 2.6 _ ISES web control library for plug-and-play RE (left) and two web pages of RE, created by ISES library – one on time dependence of temperature (right, above) and the other on the movement in two directions (right, below), measured by ISES modules

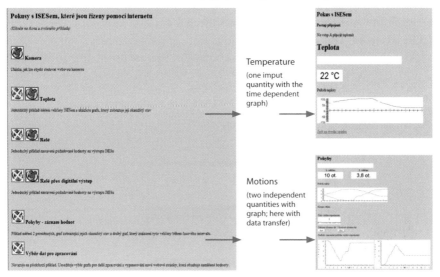

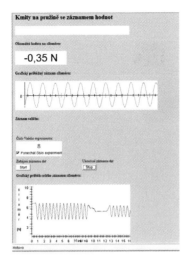

Time t (s)	Deflection x (a.u)
1E-3	-0.2773
0.013	-0.2617
0.021	-0.2344
0.033	-0.2383
0.041	-0.2031
0.053	-0.1992
0.061	-0.1797
0.071	-0.1523
0.081	-0.1211
0.091	-0.0898
0.101	-0.0742
0.111	-0.0547
0.121	-0.0273

Figure 2.8 _ Second version of web page of RE on mechanic oscillator oscillations with graph, and data output

2.4 Environment for easy building remote experiments

Easy Remote ISES

Easy Remote ISES (ER-ISES) is a graphical development environment which generates the control psc file for the finite-state machine (FSM) and the webpage code for RE controlling by a client [51]. JavaScript language is used for the compilation of code. Let us briefly introduce the development environment ER-ISES and examine how to work with it.

When we start the development environment ER-ISES, the welcome menu with photos appears, showing the versions of ISES and options that are available. (Figure 2.9).The welcome screen also allows the choice of language and the option to open the saved projects. For practical reasons the remote experiments are divided into three categories according to their complexity–starting level (a set of precompiled typical RE), basic level (the most used RE with variability and logics) and advanced level (rather complicated RE). Let's describe the main parts of the programme for each of the three RE complexity levels.

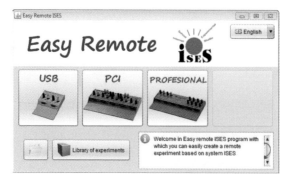

Figure 2.9 _ Welcome panel of Easy Remote ISES graphical environment development environment

Starting Level

The simplest way of ER-ISES use is the library of common experiments, from where the RE builder can choose, automatically install and start. The library is accessible by clicking on the button "Library of experiments", located on the main application panel (Figure 2.9). A window, as shown in Figure 2.10, appears. In the upper left part of the window a choice of experiments is given.

When the experiment from the library is chosen, the description of the experiment appears in the right panel along with the list of hardware

Figure 2.10 _ Experiment library window of ER ISES for starting level of compiling RE

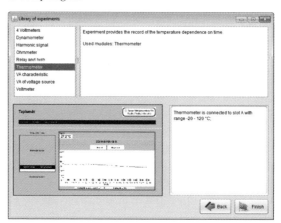

modules required. At the bottom left, there is a preview of the layout of the corresponding website (the shaded area).

If you point the mouse to the edge of the screen, the arrows for the next screen appear and the photos of devices and modules are shown. In the last part is a text with a description of the experiment. To end the compiling of the starting level remote experiment, it is only necessary to then press the "Finish" button. The programme itself generates the necessary components of the experiment (psc file and webpage code) and then returns to the start menu. At this point, the experiment is already operational, and we can start to use it. The whole procedure of RE compiling takes about 30seconds with a standard PC.

Basic Level

In the basic level of RE, the ISES version has to be specified first. The choices are given in the welcome screen (Figure 2.9). Upon choosing, the screen for the selection of measuring modules and their range adjustment is then shown (Figure 2.11).The modules are selected by using the pull-down menu on the left side of the window. When a module is selected, its photo appears in the appropriate slot of the ISES panel. Its range is adjusted using the pull-down menu, available below the module.

If the experiment uses a relay board, this option can be enabled by pressing the "Relay board" button (Figure 2.11). Once pressed, a window appears with a relay board photo, by means of which you may connect and activate 2x8 relays. The programme then asks for details of the individual relays settings. For example, we may choose manual activation with a button on a webpage, or after meeting any comparative condition (e.g. com-

Figure 2.11 _ Module selection window for basic level of RE

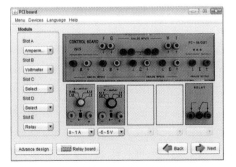

paring an arbitrary reading of any quantity and any preselected value of it, or we may compare it with the instantaneous reading of a signal generated by ISES output).

The next screen for the website design then appears (Figure 2.12). On the left side of the screen is the name of the experiment and its IP address from which the experiment will be available globally. For local operation, press the button "Run experiment locally". In this case, the experiment runs at the internal address 127.0.0.1 and can be opened only from the computer on which it was installed. Next, we can add the logo that will be displayed on the webpage and also the schematic arrangement or photo of the experiment. The other buttons are used to set the colours of the page header and footer.

The next important purpose of the screen in Figure 2.12 a is the activation of selected modules. Module names are displayed in two columns (Inputs and Outputs). The red module displayed indicates the non-activated state, while the blue one shows the already correctly activated state of the module.

To activate a module and insert the time representation of its readings (e.g. v-meter2_o), the "Add activity" button must be pressed and the catalogue of "activities" is then shown (Figure 2. 13 a). From the list we may then choose the specific activity ("Time dependence (1variable)"). After selection of the desired activity, click on the "Add to page" button and the window with parameters for the adjustment of the activity will be shown

Figure 2.12 _ IP address and website design window

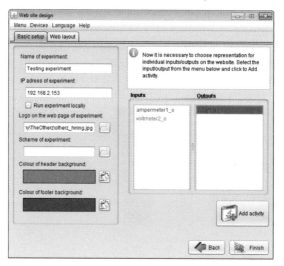

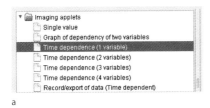

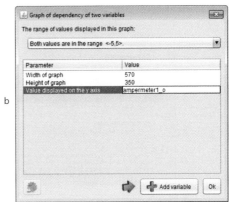

a

b

Figure 2.13 _ Window for activity choice (a) and window for activity parameters adjustment (b)

(Figure 2.13 b). Most of the parameters are already completed, and require only to be confirmed.

After activating all the modules in the list of inputs and outputs, press the "Finish" button and the programme will return to the welcome menu. From this point onwards, the RE is ready for use.

Advanced Level

The advanced level of RE compiling is based on the individual steps of the corresponding flow-chart diagram and its transformation to the psc file by using the pre-prepared control blocks of the psc library. In this way it is already possible to assemble even rather complex control logic programmes for advanced ISES RE. This method requires a certain amount of creativity and logical thinking concerning the sequence of actions of the experiment.

The advanced level of compiling the ISES RE begins at the welcome screen (Figure 2.9) by pressing the "Advanced experiment" button. The process of advanced design is similar to the basic design, where one has to select the modules and design the website. Only then, the window for advanced level appears (Figure 2.14). Its left part is a tree structure representing the control logic of the experiment. On the right side we find a list of blocks that can be inserted into a tree structure and also a list of variables.

The compiling of control logic is performed by selecting the appropriate block from the menu and then clicking at the point of the tree where the block is to be placed. Individual blocks have their own rules for inserting into the tree and may not be placed arbitrarily. Upon entering the advanced

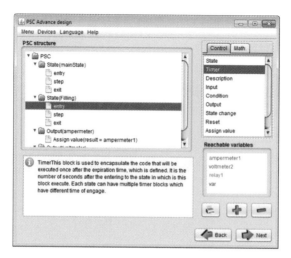

Figure 2.14 _ Advanced design window

Figure 2.15 _ Sequence of operations in one state

design, the first block "State", is already in position in the tree. This block is the main block of the entire design representing certain steps performed by the experiment. As an example, we can mention the experiment: "Water level management", which controls the liquid level between two probes (see www.ises.info).

In this experiment we distinguish various states, such as: filling, draining and waiting for the switching of the probe. The activities of the individual steps are obvious by their names. For a new experiment we will have to consider all the actions present in the experiment and then divide them into similar states before compiling the start of the controlling programme. The best way is to create a flowchart diagram of the experiment and to divide it into individual states before the start. In practice, each state has three basic parts: entry, exit and step. These parts vary in time of execution according to the code they represent. The part "entry" is always executed once, when the experiment enters the state, and similarly the "exit" part is executed on leaving the state. The middle part "step" is repeated until there is a transition to another state. The sequence of operations is shown in the following diagram (Figure 2.15). After completion of the structure by the next button we can proceed to the website design.

Other functions of ER-ISES

The development environment ER-ISES provides the storage for an unfinished draft of a RE. This makes the re-editing of the experiment very easy. This procedure may greatly shorten the time required for the design work in case of the occurrence of any errors. The saving the of the experiment (this is only possible as the last step of the webpage design) is by pressing the "Save" button in the pop-up menu. The experiments are stored in a special file with the extension .ers. The opening of a saved experiment can be made from anywhere in the pop-up menu, or by pressing the "Open" button directly on the welcome screen of the development environment. The experiment, which has been opened, is then shown in the initial step of the design. When one moves to the next steps, the data is already shown, having the exact same values as in the moment of the saving of the project.

Example of application of the ER-ISES environment

In order to demonstrate the control file and the controlling client's webpage design in the development environment ER-ISES, we present the example of the creation step-by-step. For this purpose we choose the well-known interactive experiment of electromagnetic induction of a coil rotating in a magnetic field. We are then required to choose the frequency of rotation of the coil, graphical time representation of its instantaneous voltage, data

Figure 2.16 _ Experiment arrangement of the ISES RE "Electromagnetic induction" left: coil with a driving motor and the permanent magnet; top: ISES panel with V-meter ISES, and output ISES

transfer to the client's computer, and the live view of the experiment. Let us suppose we have the hands-on ISES experiment ready where we use a DC motor for driving the coil rotation (0 – 5 V) and also have a web camera at our disposal (Figure 2.16).

Let us choose the basic level design of the controlling programme, using the following steps:

1. Start the development environment ER-ISES and select the ISES version.

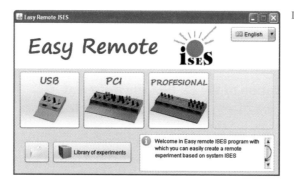

Demo _ **Figure 1**

2. Demo Figure 2 appears. We will now need to select the modules for the signal measurement of the coil voltage – ISES V-meter module in the input A. We choose the V-meter module from the drop-down menu and adjust its range.

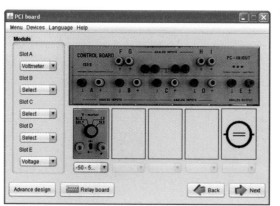

Demo _ **Figure 2**

3. For driving the motor we will select the output E as depicted in Demo
   Figure 2. We will supply the driving motor directly from the ISES board;
   and therefore, select "Voltage" as the output module. Depending on the
   motor, we may increase the output current by using the boost ER-ISES
   module. After pressing the "Next" button, the programme will ask for
   details of the setting of the output voltage.

4. The basic setup of the website is shown in Demo Figure 3. Here, we may
   enter a name for the experiment and the IP address from where the experiment is running, the logo displayed on the page or change the color
   of the page background.

Demo _ **Figure 3**

In the list of inputs we have the un-activated (shown in red) voltmeter
module connected to position A. By pressing the "Add activity" button, we
open the catalogue of activities (Demo Figure 4 a), where we activate the
module and select the representation of the voltmeter readings on the

Demo _ **Figure 4**

a

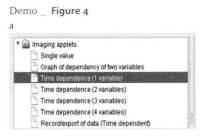

b

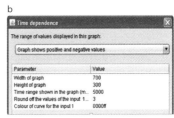

webpage. Let us select "Time dependence (1variable)". We are then asked for the adjustment of the graph's parameters (Demo Figure 4 b) and the displaying of negative voltage in the graph. The V-meter module is then activated (shown in blue when this occurs for the V-meter module).

5. We will now activate the voltage for the driving motor. In the list of outputs (Demo Figure 3) we have the reading"voltage1_i"(shown in red). We want to adjust this voltage in two ways. The first will be by pressing the button "Scrollbar", which allows a continuous adjustment of the driving voltage. Let's insert it in a similar way as in the previous case using the option "Scrollbar" in Demo 4 a. There are no parameters in the "Scrollbar" activity.

6. The other way of driving the voltage adjustment is by the five fixed selected voltages (0, 2, 3, 4 and 5 V). For this purpose we select the activity "button" (selected from the list of activities in Demo Figure 4 a). For the activity settings , select the "Setting the output voltage" button and enter the description of the button ("Stop") and voltage value 0 V ("0") (Demo Figure 5). This button will turn off the driving motor. In the same way we select four more buttons which will adjust the fixed voltages 2, 3, 4 and 5 V for the driving motor.

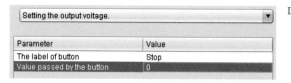

Demo _ **Figure 5**

6. The last activity we add is the capture and storing of the measured data on the client's computer. This is done by selecting the button "Record/ Export of data" (Demo Figure 4 a).

7. Let us now modify the layout of activities on the webpage. For this we use the "Web layout" tab (see Demo Figure 3) In the tree structure, we can see all the added activities (Demo Figure 6). We can use the buttons to add titles, labels and rows. By moving the activities in the tree and adding them to the rows, we can adjust the layout on the website.

8. Now we can finish compiling the remote experiment, but for safety reasons it is recommended to save it first. This is done by pressing the "Save" button in the main upper Menu. We select the name and path of the saved file and then confirm. We can then finish the experiment by pressing the "Finish" button. The experiment will now be already running and we can access it via the given IP address.

Demo _ **Figure 6**

9. The final web page layout and sample of the exported data by RE Electromagnetic induction, is shown in Demo Figure 7.

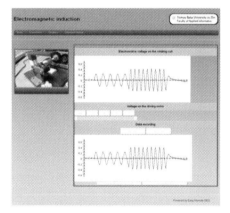

No	Time t (s)	Voltage U (V)
1	0,009	-0,4019
2	0,018	-0,4726
3	0,03	-0,538
4	0,038	-0,5743
5	0,047	-0,5688
6	0,059	-0,5198
7	0,068	-0,4436
8	0,079	-0,313
9	0,088	-0,2223
10	0,097	-0,1098
11	0,109	0,0445
12	0,118	0,1624
13	0,129	0,2985
14	0,138	0,3947
15	0,15	0,4836
16	0,159	0,5235
17	0,167	0,5688

Demo _ **Figure 7** _ Experiment arrangement of the ISES RE "Electromagnetic induction"

Chapter 3

Integrated e-Learning

3.1 Remote experiments in research-based courses

A great deal of attention worldwide has been devoted to remote experimentation and especially to remote e-laboratories [56], and these are currently undergoing a strong evolution of great interest (see recent monographs [6] [7]). Technology-enabled labs include different kinds of experiments which we can distinguish between remote, hands-on and simulated laboratories [57]. Remote laboratory experiments are further distinguished by batched experiments, interactive experiments and sensor experiments [49]. Simultaneously, a large debate is still going on which addresses the critical issue of whether remotely operated or simulation-based labs are as effective as the traditional hands-on lab format [58]. Meanwhile, online education has been strongly enhanced following the Bologna Declaration, which has contributed to shifting the learning focus from the institution to the student [59] with an emphasis on "learning by doing" using laboratories [60]. In this scenario, laboratory activities are considered essential to increase the effectiveness of teaching and learning in scientific disciplines. On the other hand, laboratory management can be resource-intensive and expensive, since it requires qualified staff, continuous equipment maintenance and evolution, so that the number of laboratories is often limited. In addition to this, the economics and logistics factors have a significant influence. For these reasons, the adoption of alternative access modes, such as remote laboratories, are desirable to schools and universities.

Remote labs in fact, can extend the capability of a conventional laboratory by increasing the number of times and places in which a student can perform experiments [61] while at the same time also extending the availability of the laboratory to several students. Moreover, they have the potential to provide affordable experimental data by sharing expensive laboratory equipment within a larger pool of users. Plug-and-play remote laboratories are used by students and researchers according to a schedule supervised by the lab manager [62]. A remote collaborative laboratory is as such, where any authorised remote user can schedule a lab session in collaboration with other users. Different kinds of users (researchers, university students, external users, etc.) must observe different management policies and scheduling rules, according to equipment features and lab manager requirements. Integration into the LMS system like MOODLE [63] and Personal Learning Environments like Graaasp [64] is expected. In a blended remote laboratory, local and remote users can collaborate when using the lab equipment (e.g. a local tutor can coach a remote group of students about a new lab technique). The integration of remote labs into the university organisation, so that they can become a tool to deliver laboratory services through the Web, is currently under way [65].

These above-mentioned requirements set forth a list of technical and organisational challenges that should be specifically addressed. How do we create and deliver easy plug-and-play HW and SW solutions for mediating the complexities of usage and creation, especially of remote real, and create and deliver all the mature and excellent solutions to the wide population of interested and would-be users of these techniques? The plug-and-play system for remote experiments at schools, both on HW side and SW side, with corresponding middleware, has never been recognised as a complex problem up until now.

Accordingly, the third point of the EU ICT-2011.8.1 project assignment (See: ftp://ftp.cordis.europa.eu/pub/fp7/docs/wp/cooperation/ict/c-wp201101_en.pdf (*iii*) *experiments for specific pedagogical contexts in primary and secondary schools and higher education, including at university level),* has not received any corresponding attention until now. In addition to the technical problems connected with remote experimentation, it is necessary to solve the methodological problems in introducing remote experimentation into the teaching and learning process. Real remote experiments which bridge the gap up until now exist in all forms of e-learning and a new strategy of education, based on the process similar to the cognition of real world by the sciences, was introduced by Integrated e-Learning (INTe-L) [66].

3.2 Basic criteria of INTe-L

The purpose of the present chapter is to offer help in the ensuing situation and to report about a new pedagogical environment, Integrated e-Learning (INTe-L) and its use in implementing remote experimentation in real education using technical means, as remote experiments and simulations [67]. At the end of the chapter the application of INTe-L at university level in distance education courses is given.(In the Appendix is a short description of the existing remote experiments built by Trnava University in Trnava. You can read more in [68]).

The present discussion about new teaching methods in physics is no longer directed towards fundamental changes in the learning processes due to the new ICT, but rather how to introduce the new techniques into everyday teaching practice by establishing the resources of e-learning, curricula, etc. Among teachers of physics exists the prevailing opinion for the necessity of change in the physics teaching strategy. In their recent paper on the physics education transformation, C. Wieman and K. Perkins [69] ask the general question *"Is there a way to teach physics that does not produce such dismal results for the typical student?"* They give the positive answer by claiming *"By using the tools of physics in their teaching, instructors can move students from mindless memorization to understanding and appreciation".* Many educators solve this problem by different approaches, many of them by increasing the role of laboratories – either real computer oriented [70], real e-laboratories across the Internet [71] or virtual laboratories and simulations [72].

We also adhere to the opinion that laboratories and simulations can deeply change the teaching of physics, but new strategies, including these new teaching tools, are needed. For that reason, we suggest the method of e-Learning, including RE experimentation, and call it Integrated e-Learning (INTe-L):

"INTe-L is the interactive strategy of teaching and learning based on the observation of real world phenomena by real e-experiments and e-simulations and on the principal features of the physic laws. It includes e-teaching tools as interactive e-textbooks and manuals and instructions which provide information and theoretical background for the understanding and quantification of observed phenomena." [66].

With a new strategy the question arises, if INTe-L solves the present difficulties in physics teaching and complies with the findings, what do physics education researchers bring to the effectiveness of the education process [73]. The prospective methods of teaching, including INTe-L, should comply with the general knowledge coming as a result of the cognitive research:

1. Students should be provided with a suitable organisational structure, based on his/her prior thinking and experience and starting from their own research results. We should not simply be pouring facts on them and not addressing the simple question of "what", but rather "why". In addition to this, any previous knowledge must be carefully checked and examined and possible misconceptions dispelled. The ultimate goal in this respect should be active thinking and active exploratory work guided by the active role of the teacher, and conditioned by the double-sided interaction of student – teacher.

2. The traditional teaching of "the rules" brings excessive amounts of new material that is far more than a typical person can process or learn. The higher cognitive load the brain is given, the less effectively it can process anything and is at the same time blocked from processing and mastering new ideas. This is one of the most well established and widely violated principles in education. Any new method that brings a remedy to the situation and maximise learning should minimise the cognitive load by minimising the amount of material presented. Presentations should be well organised and well-structured and make the link to known ideas already held by the audience.

3. The third important criterion concerns the students and public beliefs about physics education and the importance of physics for society. If the belief prevails that physics is purely abstract in nature and does not address the problems of the real world, this will deeply influence the approach towards physics as a subject and the necessity and importance of its mastery.

How does INTe-L address these three criteria? The first above mentioned criterion is met by INTe-L through its starting point, observations, irrespective if it is a traditional computer based laboratory, remote real e-laboratory across the Internet or a virtual e-laboratory [74]. The real experiments strongly support the examinations of the real world. On the other hand, the virtual laboratories or simulations support an interactive approach, employ dynamic feedback, follow a constructivist approach, provide a creative workplace, make explicit otherwise inaccessible models or phenomena, and constrain students productively [75]. The cognitive load in INTe-L is limited by supporting the individual comprehension processes offering manifold accesses to knowledge and being individually adaptive, offering significant advantages in the individual rates of the teaching progress. Traditional teaching scenarios cannot satisfy this requirement, particularly because of cognitive capacity issues. INTe-L environments meet these needs. The possibility of making abstract objects and concepts tangible by application to real and virtual laboratories demonstrates this qualitative change in education and addresses the diminishing of the cognitive load of students [76].

In the fulfilment of the third criterion, INTe-L brings the qualities and skills the students acquire through the study of physics courses for their future study and professional careers. We tried to cope with this problem in a separate paper [66]. In practical teaching it means assigning problems that are graded strictly on a final number, or that can be done by plugging

Figure 3.1 _ Illustrative photo: Students of Faculty of Applied Informatics TBU in Zlin in the physics laboratory with hands-on, virtual and remote experiments
(Photo F. Schauer, 2012)

the correct numbers into a given procedure or formula. This approach can teach students that solving physics problems is only about memorisation and coming up with a correct number. Reasoning and checking whether the answer makes sense are irrelevant. The good news is that courses that make rather modest changes to explicitly address student beliefs have avoided the usual negative shifts. Those changes include introducing the physics ideas in terms of real-world situations or devices with which the students are familiar; recasting homework and exam problems into a form in which the answer is of some obvious use rather than an abstract number; and making reasoning, sense-making, and reflecting explicit parts of in-class activities, homework, and exams. A photo of students in a physics laboratory with hands-on, virtual and remote experiments can be seen in Figure 3.1.

3.3 Example of a research-based course on AC circuits

Let us now present one of the 13RE residing in the c-laboratory at Trnava University „Transfer of energy in RLC circuits" (http://remotelab3.truni.sk) and show how it is used in the research-based INTe-L course in the part "Electromagnetism"[44]. (Further examples of 8 case studies of using RE in research-based teaching are given in Chapter 4.)

REs are used in all forms of education – lectures, seminars and laboratory exercises with the stress being on its different parts. In the lecture, it serves like a demonstration and interest inducing experiment, in seminars for theoretical calculations of properties of the circuit in question and in the laboratory for examination of the phase relations in RLC circuits and its relation to the energy transfer from the source to the load. In addition, the RE is used for project work, homework and presentations for students' competitions. Last but not least, it is used for the purpose of examinations. The material presented is the result of the experiments' use for distance students, who measured and elaborated their projects accordingly. Examples of their graphical outputs and assessment of the measurements are also included.

This RE was divided into two parts, one describing the phase relations in an RLC circuit and the second part, dealing with the energy transfer from the generator of the AC signal to the load. The webpage of the experiment is shown in Figure 3.2.

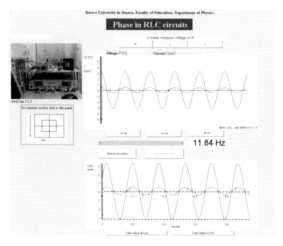

Figure 3.2 _ Web page of RE "Energy transfer in RLC circuits"

Phase in RLC circuits
The experiment is available at: http://remotelab3.truni.sk/phase_rlc.html,

Physical background
The circuit, composed of the source of harmonic electromotive force U_0 with the resistor R, capacitor C and inductor L in Figure 3.3 may be described by the differential equation for current on circuit $I = I(t)$

$$L\frac{d^2 I}{dt^2} + R\frac{dI}{dt} + \frac{1}{C}I = -\omega U_0 \sin(\omega t). \tag{3.1}$$

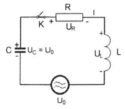

Figure 3.3 _ Scheme of RLC circuit with AC source

This equation has a solution in the form

$$I = I_0 \sin(\omega t + \varphi), \tag{3.2}$$

where

$$I_0 = \frac{U_0}{\sqrt{R^2 + \left(\omega L - \frac{1}{\omega C}\right)^2}}, \tag{3.3}$$

with the initial phase (the phase shift of the current I against voltage U)

$$tg\, \varphi = \frac{\frac{1}{\omega C} - \omega L}{R}.$$

It is evident from (3.2) and (3.3) that the amplitude of the current I_0 and the initial phase φ depend both on the values of the components R, L and C, and on the angular frequency $\omega = 2\pi f$.

Energy transfer in RLC circuits

http://remotelab3.truni.sk/transfer_energy.htm

Physical background

The main purpose of the RE is to show the frequency dependent energy transfer in the series RLC circuit from the source to the load resistor defined by

$$P_{av} = \frac{1}{T}\int_0^T Ri^2 dt, \tag{3.4}$$

where $i(t)$ is the instantaneous current and the term Ri^2 the instantaneous power. The equation (3.4) may be simplified to the form

$$P_{av} = \frac{U_0^2 b}{\omega_{res}^2 L} \cdot \frac{1}{\dfrac{b^2}{\omega_{res}^2} + \left(\dfrac{\omega}{\omega_{res}} - \dfrac{\omega_{res}}{\omega}\right)^2},$$

(3.5)

where Uo is the amplitude of the source voltage, ω_{res} is the resonance frequency and the damping coefficient $b = R/2L$. Figure 3.4 describes the frequency dependence of the energy transfer in a series RLC resonant circuit, calculated using amplitude and phase characteristics and equation (3.5). The final results of the measurements are shown in Table 3.1.

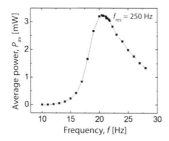

Figure 3.4 _ Energy transfer in a series RLC resonant circuit

Quantity Value	
Resonance frequency	f_{res} = 20.6 Hz
Amplitude of current in resonance	I_{max} = 2.3 mA
Maximum of average power	P_{str} = 3.2 mW

Table 3.1 _ Final results of the measurements "Energy transfer in series RLC circuit"

Let us now show the accompanying materials for INTe-L teaching for lectures in Figure 3.5, seminars in Figure 3.6 and a laboratory exercise in Figure 3.7 available at LMS MOODLE.

22.3 AC current in AC circuits

22.8
Electric circuit with harmonic electric current, is in general composed of three idealized elements: resistor, capacitor and inductor, with their quantities of resistance R, capacitance C and inductance L. Total "resistance" of the circuit composed of the in series of these elements connected circuit is called impedance Z

$$Z = \left[R^2 + \left(\omega L - \frac{1}{\omega C} \right)^2 \right]^{\frac{1}{2}}.$$
 (22.21)

If the circuit is composed of real elements we speak about capacitor and inductor described not only by their capacitance C and inductance L, but also by certain resistance R.

22.9
In the RLC circuit the harmonic current is shifted with respect to the harmonic voltage by the phase shift φ

$$tg\ \varphi = \frac{\omega L - \frac{1}{\omega C}}{R}.$$
 (22.22)

Fig 22.11.Phase shift of the harmoinc current with respect to the harmonic voltage in RLC circuit

Figure 3.5 _ INTe-L seminar accompanying material with RE, on MOODLE

Seminar 7 - Electromagnetic field

Try simulation – to Exercise 1. Fendt	To Exercise 2. and 3. Wire with the current, moving in the magnetic field II 29.2
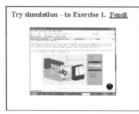	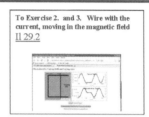

1. A flat circular coil with N = 50 windings and the radius r = 0,1m is uniformly revolving with the frequency f = 30 Hz round the axis, perpendicular to the vector of magnetic field B = 10^{-3} T of the homogeneous magnetic field. **Find the time dependence of both the magnetic flux and the electromotive voltage in the coil.**
2. Horizontally oriented metal wire with lenth L = 2 m starts to fall with zero starting velocity in gravitation field of the Earth. **Determine the voltage between ends of the wire in time t = 3 s,** if the wire is oriented at the beginning of the motion in the direction N-S (W-E).

3. Metallic rod of the lengthe 2 m is moving with the constant velocity v = 2 ms^{-1} perpendicularly with long straight wire with the current I = 40 A at a distance of 30 cm. **Calculate the electromotive voltage induced in the rod.**

Transfer of energy in RLC circuit Try resonance and phase shift first – here.	*Modelling environment for AC cirucuits – here*
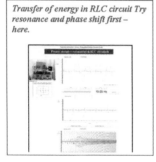	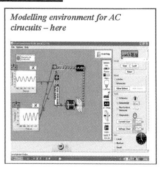

Figure 3.6 _ INTe-L seminar accompanying material with RE, on MOODLE

Examples of students' graphical output

The main goal and assignment was to become acquainted with the current amplitude $I_o(f)$ (Figure 3.7 a) and phase shift frequency $\varphi(f)$ (Figure 3.7 b) characteristics of the RLC circuit (a) theory; b) experiments).

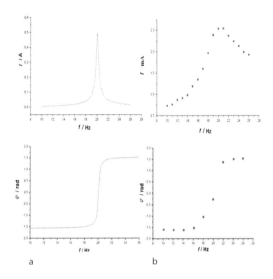

Figure 3.7 _ Examples of students' reports current amplitude I_o on frequency f (up), phase φ on frequency f (down), theoretical (a), experimental (b)

Example of a students' assessment of the measurements

"This experiment helps to transfer theoretical knowledge to practical skills irrespective of the fact that it is a remote experiment. The "remoteness" sometimes causes some trouble with the assignment, as in hands-on laboratories the instructor is always present and ready to explain any misconceptions. I want to stress however, that the advantages outweigh the disadvantages, because the other alternative would be to have no experiments at all."

"The remote experiment I worked on is a high quality tool for understanding the subject matter in question."

"For the remote experiment execution, it was necessary to understand the physical basis of the observed phenomena in RLC circuits (resonance, changing amplitudes, phase shift, etc.) and the mastering of the mathematical background was also absolutely necessary. A very great acquisition was the use of modern ICT for experimentation and signal capturing, which are a great help and make it easier to understand the phenomena in RLC circuits."

Another eight examples of the use of RE in research based INTe-L courses are given in the next chapter.

Chapter 4

**Remote experiments
in research-based education**

The present state of the ICT branch now enables remote access and control of physical experiments with real-world experimental data transfer. Consequently, the gap which existed in e-Learning, i.e. lack of remote experimentation, was removed. Simultaneously, the possibility to create a new generation of e-Learning with remote experimentation occurred and Integrated e-Learning (INTe-L) was suggested [77][78]. The history of the development of both e-experimentations and INTe-L strategy was dealt with in Chapter 3. In the present chapter we intend to show the changes INTe-L has brought into the education process with respect to research-based teaching. In addition, we want to show the technical means which enable simple ways of searching and approaching the most suitable experimental tools and concomitant materials. Modern teaching systems serve this purpose, known under abbreviation as LMS (Learning Management System), and their higher RLMS, described in Chapter 5.

Using these tools, in our case MOODLE and in near future REMLABET, we will show next the practical exploitation of teaching strategy INTe-L with RE for physics teaching at secondary and university levels.

4.1 MOODLE with remote experiments in research-based teaching

LMS system MOODLE belongs to the most spread-free accessible teaching software. The acronym MOODLE comes from Modular Object Oriented Dynamic Learning Environment designed by Martin Dougiamas from New Zealand during postdoctoral studies, at present active in Curtin University in Australia [79]. MOODLE has now more than one million applications all over the world. It is currently being continuously maintained and upgraded (more detail on http://moodle.org [80]).

Once we have study materials in electronic form, MOODLE requires only a server, IP address, and easy installation. It is then accessible via an Internet standard web browser, so after downloading the study materials and the signing in of all students, it is immediately ready for use in all forms of teaching and learning. MOODLE can then be used as a communication and management tool for the educational process and as a repository for all course study materials. Specifically, it registers the following activities: signing-in of students, controlling their activities, following-up of their progress in acquired knowledge, how efficient they are becoming in all types of education activities and their grading. It also enables the submission of homework, projects and last but not least enables a rich communication between the student–teacher–student. As a module, we use software clickers in MOODLE on the basis of client-server communication IPAL [38].

In connection with the INTe-L strategy a question arose: *"Why do we not use the MOODLE LMS system with all its advantages for all three basic components of INTe-L, i.e. remote experiments (REs), virtual experiments (VEs) and e-textbook for physics teaching?"* The examples presented next will show how the idea succeeded.

The beginning of teaching by means of INTe-L in a basic course of physics may be dated from the school year 2007/2008 when the first courses of physics for informatics students Physics 0 – Mechanics, oscillation and waves, Physics 1 – Electricity, magnetism and electromagnetisms, Physics 2–Quantum physics and Solid state physics were accessible on the server www.vyuka.fai.utb.cz.

Samples from the course are presented in Figures 4.1– Figure 4.4. Every enrolled student on the course communicates with all the teachers on the course, in all forms of teaching and other activities. In all forms of teaching interactive remote experiments are used (predominantly from the servers

Figure 4.1 _ Front page of the course Physics 0

Figure 4.2 _ Front page of the course Physics I

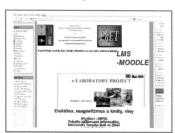

Figure 4.3 _ Front page of the course Physics II

Figure 4.4 _ Detail of the course Physics 0

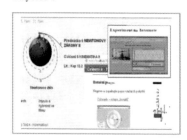

www.ises.info and http://kf.truni.sk/ remotelab, or just directly http://re-moelabN.truni.sk, where N = 1, 2, ..., 13). An indispensable part of research-based teaching is the use of simulations as virtual experiments. Above, you will see screenshots taken from the webpage from the University of Colorado (http://phet.colorado.edu) and interactive Physlet from the workshop of W. Christiana a M. Belloniho [81].

On the basis of the research-based education of dynamic of mass point, using ICT means provided to the students via MOODLE, let us now show how to use remote experiments in individual forms of teaching: i.e. lecture, seminar, laboratory exercise, project work, homework and examination with the use of INTe-L.

A) Lecture – RE may be used in a lecture as an introduction, to motivate students and draw their attention, with the advantage of using a large interactive board with the possibility of graphical comments. Figure 4.5 presents the front page of the lecture in MOODLE with real and virtual experiments "Mechanical oscillator" and a moving body pulled by a weight on a horizontal substrate, which we show at the beginning of the lecture. For example, the lecturer asks the following motivating questions when discussing the

Figure 4.5 _ Sample illustration from the course in MOODLE – lecture

Newton's Laws of Motion

Experiment on the Internet:
Damped and driven oscillator

Questions:

1. Why does the body on the spring move periodically?
2. Why does this periodic motion stop after some time?

Sir Isaac Newton, 1643 – 1727

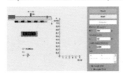

Try to solve this problem:
http://www.walter-fendt.de/ph6en/newtonlaw2_en.htm

mass and spring oscillator: "What is the cause of the motion?", "Why is the weight moving periodically?", "Why does the motion cease after a certain time?" and thus following these questions a discussion ensues. Examples of oscillatory motion serve as an example for discussing the causes of motion, their origin and their properties. It is useful to use simple simulation, e.g. Forced body motion using a simulation body and weight (http://www.walter-fendt.de/ph14e/n2law.htm), which is important for understanding the causes of motion. At this point , we may also mention the already acquired knowledge from kinematics and stress the importance of the acceleration as the second derivative of the instantaneous position. During the lecture, all Newton's Laws of Dynamics are presented in a similar way using several remote experiments.

After presenting Newton's Laws of Dynamics, we may obtain feedback of the conceptual understanding by means of software clickers (Figure 4.6). In our case we used the software module for MOODLE 2.6 designed by Dr. Wiliam Junkin z Eckert College, St. Peterburg, Florida, USA [38]. As a matter of fact, the SW clickers are the remote experiment on the communication scheme server-client with multiple clients and accessible across arbitrary communicators equipped with the Internet.

Students answer the following question: "Four elevators lift the weight of the different mass m with various forces F to the equal height h. Which elevator wins the race?" (Figure 4.7 left). The answers are shown in the column diagram (Figure 4.7 right) available to all students. The feedback provides the teacher with speedy information regarding understanding of

Figure 4.6 _ View of the teaching process with wireless clickers (IPAL modul in MOODLE 2.6) by means of student's mobile devices, Faculty of Applied Informatics, Tomas Bata University in Zlín, (photo P. Dvořák)

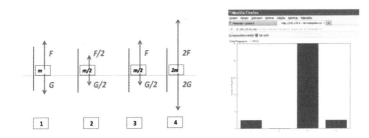

Figure 4.7 _ L eft: test question from dynamics, right: students' answers

the subject matter taught on Newtons' Law of Dynamics. The teacher may then analyse the answers, explaining the correct answer in the form of an open discussion with the class. In the extreme case, if and when a large deviation from the correct answer occurs, the possibility of other methodological procedure in explaining any misconceptions may be used.

B) Seminar – All seminars take place in computer rooms, where every student has his/her own computer connected to the Internet, enabling him/her to work at his/her own pace. Demonstration examples and research and problem solving tasks are available in seminars. During seminars, students, under supervision of the teacher, solve assigned typical tasks,

Figure 4.8 _ Example of the seminar task – dynamics

Seminar – Newton's Laws

3. In the simulation, the body of the mass $M = 0.125$ kg is pulled on the base by the weight of the second body of the mass $m = 5.2$ g. The motion is described by the equation of path $x = 0.1\,t^2$, where x is the coordinate and t is the time. Find the coefficient of friction of the body and the base.

(Follow the comment to the previous problem.)

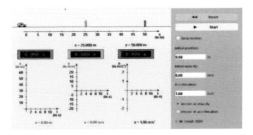

predominantly those that caused problems during home preparation. In addition, the class works with simulations and remote experiments that are explanatory and instrumental to solve problems. An example of the assignment of a task for the seminar is shown in Figure 4.8.

In the same manner, students are examined across the Internet by tests on mastering the subject matter of the previous seminar, which is the only evaluation of the student's activity. The results of individual tests together with the other evaluated activities from the course are gathered and summarised in the LMS MOODLE environment. Each student is acquainted with the course rules, their running fulfilment and the total mark, which is accessible for each student.

C) Laboratory exercise – It is often the case that the experiments are in the laboratory in either hands-on, virtual or remote forms. We leave it up to the student which form of experiment he/she chooses for the assigned topic. For the popularity of remote experiments one student makes the following statement:

"I choose remote experiments because I can measure in my pyjamas."

For the laboratory exercise, for example with distance students who cannot access the laboratory, the students measure and use the remote real or virtual experiments to their advantage.

D) Project – The course Physics 0 – Mechanics serves for the repetition of the basis of the secondary school curriculum. It is based on the fulfilment of individual project elaboration, which the student chooses from a database of around 130 different simulations with data output and remote experiments, so that each student elaborates the problem from a different part of the delivered course. Simulations or REs are used as a source of the data, which after processing and evaluation, are compared with the modelling results, based on the theory, which gives a basis for constructive discussion. The teacher is in contact and may communicate with a student across MOODLE and may return the project for amendment and improvement of the mark. Figure 4.9 depicts a sample of MOODLE unit of project download.

Let us present the sample of project work "Verify the maximum height of a golf ball by measurement and calculation ", whose goal was to compare data outputs from the Christian and Belloni simulation [81] accessible from http://zamestnanci.fai.utb.cz/~schauer/contents with the theoretical

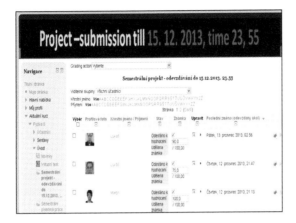

Figure 4.9 _ Sample of MOODLE unit of semester project download

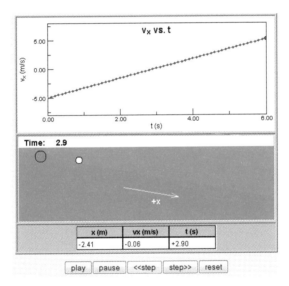

Figure 4.10 _ Project "Verify correctness of the simulation – Golf ball motion"

model designed by the student himself. It is the model motion of the ball on an inclined golf pitch with a hole in it (Figure 4.10). The aim is to determine theoretically the initial velocity of the ball to hit the hole. In the lower part of Figure 4.10 this situation with the ball, hole and inclination and the graphical representation of the equation of the velocity (upper part). is schematically depicted. The student has to determine the initial velocity required to reach the hole with zero velocity.

The graph in Figure 4.11 depicts the equation of the path determined both theoretically (black points and full line) and the data collected from the simulations (red points).

Up until now, gained experience shows [82][83][84] that students, irrespective of the efforts with project elaboration, praise this form of study and the knowledge acquired.

During elaboration of the project and the practical activities during experimentation and study of literature, the students come to realise just how many skills they have to master, and in return, how much depth of knowledge and understanding of physical laws they gain. These facts are confirmed by a sample sentence from the student's project conclusions on the assignment: "Verify by measurements and calculation the maximum height of a golf ball":

"The work on the project was very comfortable. I had to repeat a great deal of subject matter for future use and also for the semester written exam. Even though the problem was quite easy, introducing many simplifications, I am glad that I was given the chance to work on exactly this project. As a whole, I appreciate the project work as a very rewarding activity and also a step forward in my study at this school."

The total student's semester evaluation consists of several activities that enter into the final assessment (o – 100 points) with various weights: seminary tests (o – 30 points), closing written examination (o – 30 points), research-based project (o – 30 points) and teachers' opinion (o – 10 points).

Figure 4.11 _ Time dependence of the position of the ball – "theoretical" (black points) and experimental data (red points)

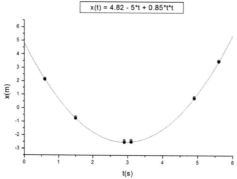

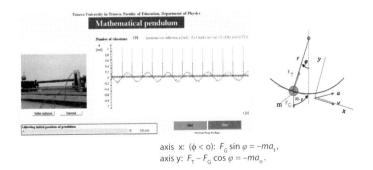

Figure 4.12 _ Illustration of the student's semester final evaluation of the course Physics 0 – Mechanics in LMS MOODLE

Figure 4.12 shows the illustration of the semester final evaluation from LMS MOODLE for one of the previous courses.

E) Homework – LMS MOODLE provides student with assignments of exercises and tasks for self-study of a theoretical background.

F) Examination and testing – The examining teacher, through discussion with the student in regard to the results of both the remote experiments and simulation, may easily assess their knowledge of the physics laws in question. As an example, we may mention the topic of the dynamics using RE "Simple pendulum" (http://remotelab5.truni.sk), from Figure 4.13 (left), with the representation of its dynamics model (right).

Figure 4.13 _ Example of examination question for Newton's law of Dynamics with RE "Simple pendulum", http://remotelab5.truni.sk

axis x: ($\phi < 0$): $F_G \sin \varphi = -ma_t$,
axis y: $F_T - F_G \cos \varphi = -ma_n$.

LMS MOODLE also provides rich testing possibilities, which may provide both the teacher and the student very effective feedback for evaluating pedagogical process effectiveness, students' opinions, and approaches. As an example, let us give the results of the standardized COLLES (Constructivist On-Line Learning Environment Survey) research, which was part of the MOODLE environment during the years 2008-2010.The test consisted of 24 test items in regard to the importance of the course, providing space for thinking and interactivity, support from school peers and the teacher and the possibility to make interpretation easier. The results of the test from 110 respondents of the research-based course Physics 0 using the strategy of INTe-L, are represented in the diagrams shown in Figure 4.14, where the cyan columns indicate the student's expectation and where the brown ones show the course reality.

The analysis shows that the students consider the course of medium importance, the same for effective thinking. From the graphs, we can also see that expectations differ from a real situation in the course. The students appreciated the teachers' help, reflecting also the extensive use of ICT support and INTe-L strategy.

As we have already mentioned, the substantial advantage of MOODLE is the possibility of a continuous check of the students' results in the course, which are being continuously renewed.

Another advantage of compiled and used courses in LMS MOODLE, which up to now has not been mentioned, is their sharing with other educational institutions. For example, students of Trnava University in Trnava appreciated the advantages of the possibility of using MOODLE courses in the fields of Physics and Quantum Physics with all their features including

Figure 4.14 _ Evaluation of the standart COLLES test in LMS MOODLE

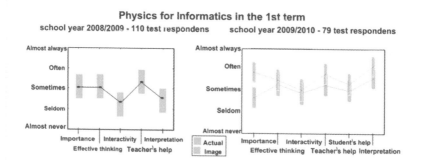

materials for lectures, seminars and laboratory exercises, as well as project work.

After we have acquainted the reader with ICT building blocks in the LMS environment MOODLE, let us present some examples of research-oriented teaching. As described in Chapter 3, the main focus of research-oriented teaching is the independent experimental and creative activity of each subject taught. By filling in the missing term into standard e-Learning in the form of RE, a real possibility of the creation of a higher form of e-Learning for research-oriented teaching and learning is available. As our contribution to this trend, we introduced that higher form of e-Learning – INTe-L [78].

Let us present the possibilities offered by INTe-L in research-oriented teaching on seven examples of units: Environmental monitoring Motion in the gravitational field, Oscillations, Dissipative processes, Electric circuits, Electromagnetic induction and Photon theory. We intend to show to which extent the units presented fulfil the basic criteria for research-based teaching formulated in Chapter 1 and also to give motivation for those who are possibly interested as well as followers of modern forms of education. In the following text, we will not present flawless instructions but rather prefer to show the responses of students and examples of their laboratory work or projects without any major amendments. (Note to the reader – to easily distinguish the students' work examples we have placed corresponding text in blue frame with a stronger staining on the left side of the picture.).

4.2 Research-based teaching – "Environmental monitoring"

Let us firstly show a simple form of RE with data collection and storing, without a possibility of influencing the conditions of the measurements (in literature such experiments are denoted as data loggers) – "Environmental Monitoring" in Trnava, Slovakia (http://remotelab1.truni.sk).

The WWW introductory open page is shown in Figure 4.15. The RE collects basic environmental data – such as temperature, pressure and sun radiation intensity.

The role of the researcher is only to choose the required quantity i.e., the temperature, pressure or sun radiation intensity and the requested time interval. After displaying this data in the graph and it's transfer to the clients' computer, one can process and evaluate the data in a common

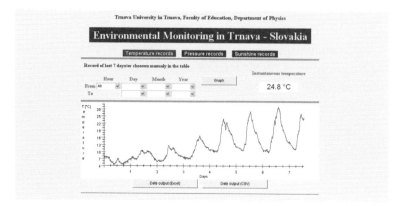

Figure 4.15 _ Introduction web page of RE "Environmental monitoring in Trnava" web: http://remotelab1.truni.sk

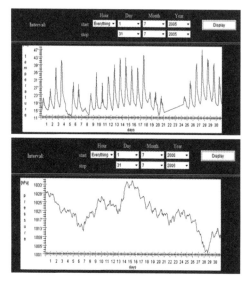

Figure 4.16 _ Graphical record of the temperature (up) and the pressure (down) (January 2005) from RE "Environmental monitoring in Prague" web: http://www.ises.info/index.php/sk/laboratory/experiment/meteorological-station-in-prague

way. Using this data, students may learn to evaluate data files, and how to present them both in tabular and graphical form, and thus gain important knowledge of the real world – from meteorology and the environment. An example of graphical representation of the temperature and the pressure data (January 2005) is shown in Figure 4.16. For comparison, the student can use the data from another meteorological station, e.g. Prague and Porto:

- http://www.ises.info/index.php/sk/laboratory/experiment/meteorological-station-in-prague,
- http://experimenta.fe.up.pt/estacaometeorologica/historico.php?lang=en).

Data processing may contribute to the teaching of mathematics in the thematic unit functions of one variable, finding dependent and independent variables and finding extremes of function as a starting point of calculus. Let us show data processing in a laboratory exercise or a project.

A) Laboratory exercise/Project – The assignment is to collect, process and evaluate data gained from the environmental meteorological station and thus train manipulation with bulky data, it's choice and filtration, arrangement into tables and graphs and its evaluation according to the assignment. Generally we have to avoid very often used "recipe" type of work. The student solves the following steps of the general assignment step-by-step at his/her own pace:

Go through the experiment and transfer the temperature data of a selected time interval from a year from RE into your computer (e.g. one day or one week), and arrange it into the table and the graph in a graphical environment as Origin or Excel processors. Repeat this procedure for the following two years.

At this moment, the student discovers (data are stored in RE every 10 s) that he/she has to deal with an enormously bulky file (Table 4.1), whose processing is of the utmost difficulty. In order to cope with this problem he/she has to select data of interest (e.g. every hour, maxima or minima, the steepest decrease or increase, etc.) through filtering. The results are

Table 4.1 _ Measurement data of the day .... and its processing

Measurement order	Time τ (h)	Temperature t (°C)
1		
....		
n		
		$\overline{t_1}$ = ... °C

then arranged into a small and simple table (Table 4.2). It is then relatively straightforward to plot the new data in a simple graph.

Table 4.2 _ The data and processing of the measurements on date..........., time .... name ....

Number	Time τ (s, h)	Temperature t_i (°C)	$\Delta t_i \Delta t_i^2 \sigma$ (°C)(°C ^{2})(°C)
1			
n			
		$\bar{t}$ = ... °C	

The student continues in general assignment:

Give your detailed observation of the time record of the temperature and attempt its short first hand description. For the time interval in question determine the average day temperature(s) (or week) and plot the corresponding curve into an existing graph.

The student may next combine all the assigned e.g. three years' results into one graph where dependent and independent variables are plotted in basic SI units and are distinguished by different colours (or with different style, e.g. dot, dash and dot, etc.).

As a next choice, the result of the temperature in a certain time interval may be evaluated and written in the form

$$t = (\bar{t} \pm \sigma)\,°C\,,$$

where σ is the standard deviation, defined by the expression

$$\sigma_{t_i} = \sqrt{\frac{\sum_{i=1}^{n}(t_i - \bar{t})^2}{n(n-1)}}\,,$$

where t_i is the instantaneous, i^{th}, value of the temperature measured at the time τ_i, $\bar{t}$ is the average temperature and n is the total number of measured points.

The students' results are shown in Figure 4.17. For more advanced students may we suggest more demanding tasks, e.g.

Time dependence of the temperature and pressure, July 13, 2005

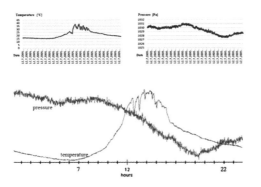

Discussion:

In the introduction, I claimed high pressure means good and sunny weather and, on the opposite, low pressure cloudy or rainy weather. After plotting both curves into one graph it was clear to me this idea was correct. But under certain conditions unexpected situations may occur changing this relation. The temperature was decreasing from 1 AM to 7 AM, than it had an increasing tendency to 3 PM when it again started to decrease till 7 PM. But the pressure had a decreasing tendency during the whole day till 7 PM when it started to increase. After 10 PM we can observe a decrease in the temperature and an increase in the pressure.

Conclusion:

I chose the data to saturate my curiosity, as I wanted to test the validity of the commonly accepted knowledge between the atmospheric pressure and the weather. Thanks to this work I had to gather data of the pressure, the temperature, but also the wheatear. The results and gained information took me by surprise as before I considered the relation between the pressure and the temperature universal.

Figure 4.17 _ Sample of the students' project "Study of temporal dependence of the temperature and the pressure"

Determine the result of the sum of the differences among the instantaneous and average temperature and verify the result by calculation,

Correlation coefficient calculation

For the calculation of Pearson correlation coefficient for the pressure and the temperature I used Microsoft Excel 2007 with the function Data analysis, Correlation. Table 1 presents the data gained

Table 1

Date and time	Temperature	Pressure	Correlation
13.7.2005 0:00:23	18,0948	1029	-0,214742290
13.7.2005 1:00:23	17,3425	1028	-0,204742290
13.7.2005 2:00:23	16,5901	1028	-0,214742290
13.7.2005 3:00:23	15,9746	1028	0,302241854
13.7.2005 4:00:23	15,4958	1028	0,316465205
13.7.2005 5:00:23	14,9487	1028	0,449974037
13.7.2005 6:00:23	14,7435	1028	-0,361125371
13.7.2005 7:00:24	15,0855	1028	0,275276078
13.7.2005 8:00:23	15,9746	1028	-0,069714887
13.7.2005 9:01:23	17,6160	1028	-0,001903360
13.7.2005 10:00:23	20,1466	1028	-0,527383440
13.7.2005 11:00:23	25,1392	1027	-0,540219762
13.7.2005 12:00:23	33,0044	1027	-0,745337002
13.7.2005 13:00:23	36,2189	1027	0,037146953
13.7.2005 14:00:23	38,1339	1026	-0,354173798
13.7.2005 15:00:23	34,3723	1026	0,498074347
13.7.2005 16:01:23	32,8676	1025	0,515524232
13.7.2005 17:00:23	26,6439	1025	0,260116370
13.7.2005 18:00:23	24,8657	1024	0,732362124
13.7.2005 19:01:23	23,5662	1023	0,544708857
13.7.2005 20:00:23	21,9932	1024	-0,733340502
13.7.2005 21:00:23	21,0357	1024	-0,606006271
13.7.2005 22:00:23	20,2149	1025	-0,740788010
13.7.2005 23:00:24	19,9414	1024	-0,030912563
13.7.2005 23:59:23	18,9155	1025	-0,501514548

The average value of the correlation coefficient is -0.39545. It indicates that on July 13, 2005, there was no significant correlation between the observed quantities, the pressure and the temperature.

Figure 4.18 _ Sample of students' project "Finding Pearson's correlation coefficient of the pressure and the temperature

or:

Determine the largest decrease or increase of the temperature in unit time (the increase or decrease in unit time is called gradient),

or:

Compare the measured data of the pressure and the sun radiation intensity

*(or the temperature and the sun radiation intensity, or the temperature and
the pressure) and try to find their correlation.*

An example of correlation processing of the measured date and finding
Pearson's correlation coefficient of the pressure and the temperature for
13. 07. 2005 taken from the students' project is shown in Figure 4.18.

Closing the first unit, we may now add that the remote experiment "Environmental Monitoring" may be used with an advantage in many other
situations. It depends on the activity and imagination of the teacher and
the data processing ability of the student. As the number of meteorological
stations accessible across the Internet is constantly increasing, the possibilities of their exploitation is increasing as well, so that comparison and
rich co-operation among different school teams is abundant (e.g. on Porto
– Prague – Trnava).

The project titled "Measurements of the air temperature in different
European Union countries" was dealt with in physics in Slovak primary
schools using remote meteorological stations in respective countries. [85]
[86][87]. Successive pedagogical research proved [87] that the teaching of
respective subjects with research-oriented approach brought better results
both in pupils' knowledge and skills with analytical abilities and predominantly increased their understanding and skills of working with graphs.

4.3 Research-based teaching – "Motion in the gravitational field"

The experimental basis for the research-oriented activities for motion in
the gravitational field is RE "Free fall" (http://remotelab4.truni.sk/position.
html). This experiment uses the motion of a permanent magnet in two identical glass tubes with equidistant pick-up coils, one filled with atmospheric
air, and the other empty. (Figure 4.19) Surprisingly, this simple experiment
enables a couple of analyses and methodical approaches from both the
point of view of motion kinematics, its dynamics and energy, including dissipative forces. Let us show the initial data the experiment provides and
the information it may give. In the table shown in Figure 4.19 the first two
columns illustrate time dependence of the instantaneous position of the
moving magnet in the tube.

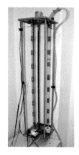

The passage times t of the magnet through the centres of the coils situated at y

t [s]	y [m]	t² [s²]	2y [m]
0.000	0.000	0.000	0.000
0.109	0.058	0.012	0.016
0.179	0.160	0.032	0.032
0.230	0.260	0.053	0.520
0.272	0.361	0.073	0.721
0.309	0.459	0.096	0.919
0.342	0.560	0.117	1.120
0.374	0.659	0.140	1.317
0.402	0.761	0.162	1.523

t – The passage times of the magnet through the centre of the respective coil

y – Possition of the centre of the coil

Figure 4.19 _ RE "Free fall", two glass tubes (with atmospheric pressure and evacuated) with pick-up coils and magnetic vessel (yellow) for magnet lifting to the starting position (left); data (right)

In Figure 4.20 a we see both the measured points and second order fitting polynomial $y = Ct^2$ with initial conditions $y = 0$ m and $v_y = 0$ m.s⁻¹ for $t = 0$ s (see inset in Figure 4.20) we obtain the value of coefficient $C = 4.89$ and comparing with the later derived equation of the path of free fall $y = (1/2) gt^2$ (see equation 4.1) value for acceleration of gravity $g_1 = 9.78$ m.s⁻². We advise students to use another possibility of the evaluation of the measured data by plotting the measured dependence in axes y vs. t^2 (Figure 4.20 b). The slope of the resulting straight line gives the acceleration of gravity $g_2 = 9.71$ ms⁻².

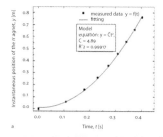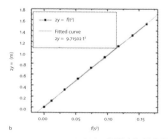

Figure 4.20 _ RE "Free fall" (a) time dependence of the instantaneous position y, (b) dependence of the instantaneous position 2y on time squared t^2

Next, let us show the use of the RE in research-based teaching within INTe-L in all forms of the topical unit Mechanics.

A) Lecture – In a lecture, we start with this experiment for motivating a discussion about the technical design of the experiment and all influences decisive for the experiment (dissipative friction forces, starting with non-zero velocity, etc.).

According to the taught part of the lecture, we use three different approaches, kinematic, dynamic or energetic.

Kinematic approach – supposing constant acceleration and from this the following equations of the path and velocity

$$y = \frac{1}{2}gt^2, \ v = gt, \tag{4.1}$$

we may obtain the acceleration of the gravity. To the students surprise we obtain its value that differs from a generally expected value. It gives the possibility for discussion about the causes of the observed difference. After repeating the experiment in the empty tube we get better agreement with a tabulated value and explain the reason to the students. We leave further reasoning in regard to the way of including dissipative forces to their own research and simply point to the modelling tools available.

Dynamic approach – the basis of Dynamic approach for solving general free fall problems is the Newton equation of dynamics for the acceleration a of the mass point of the mass m under acting of forces

$$a = \frac{\sum F}{m}, \tag{4.2}$$

where $\sum F$ is the vector sum of all acting forces. In case of the absence of dissipative forces in one dimensional motion in the direction of positive axis y, we may write

$$\frac{d^2 y}{dt^2} = g, \tag{4.3}$$

where g is the acceleration of gravity. After successive integrations with the use of the initial conditions $t = 0$ s, $y = y_0 = 0$ m, $v = v_0 = 0$ m.s^{-1} we obtain the equation of the path of the falling magnet

$$y = \frac{1}{2}gt^2. \tag{4.4}$$

In more general cases the dissipative forces F_v exist, which increase with increasing velocity and point against it. Then the situation may occur, when the gravity force equals to the dissipative force. When the dissipative force depends on the velocity value linearly, with the proportionality constant k_1, $F_v = k_1 v$, then the final v_∞ velocity is

$$mg = k_1 v_\infty. \tag{4.5}$$

Then general equation of motion (4.2) is

$$m\frac{d^2 y}{dt^2} = mg - k_1 v, \tag{4.6}$$

with the solution for the instantaneous velocity

$$v(t) = v_\infty \left(1 - e^{\frac{-gt}{v_\infty}}\right), \tag{4.7}$$

the instantaneous position

$$y(t) = y_0 + v_\infty t + \frac{v_\infty^2}{g}\left(e^{-\frac{gt}{v_\infty}} - 1\right). \tag{4.8}$$

For the case of higher velocities the dissipative forces depend on the velocity quadratically $F_v = k_2 v^2$, the equation of motion (4.2) is

$$m\frac{d^2 y}{dt^2} = mg - k_2 v^2 \tag{4.9}$$

with the solution for the instantaneous velocity

$$v(t) = -\sqrt{\frac{k_2}{mg}}\,\text{tgh}\left(\frac{k_2 g}{m}t\right) \tag{4.10}$$

and for the instantaneous position

$$y(t) = y_0 + \frac{m}{k_2}\ln\cosh\left(\sqrt{\frac{k_2 g}{m}}t\right). \tag{4.11}$$

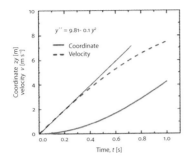

Figure 4.21 _ Modeling of position $y(t)$ and velocity $v(t)$ in dissipative media

The examples of numerical solutions are in Figure 4.21 for the high velocities (equations (4.10) and (4.11)) for the parameters k_2 = 1 m^{-1} a m = 10 kg.

Energetic point of view – usefulness of energy approach of this remote experiment may be shown in situations where we are interested in the velocity without dissipation. Then, a simple equation of the conservation of the mechanical energy, i.e. sum of the kinetic E_k and potential gravitation energy E_{pgrav} is constant, so for their increase is

$$\Delta E_k + \Delta E_{pgrav} = 0 \qquad\qquad (4.12)$$

or

$$\frac{1}{2}v^2 = gy$$

and finally we get equation $v = v(y)$

$$v(y) = \sqrt{2gy} \qquad\qquad (4.13)$$

accessible for the measurement.

B) Seminar – In the seminar, we use both virtual and real remote experiments. According to the maturity and level of the students, different assignments may be given for elucidation of various types of motion in the gravitational field. For lower level student we study motion without dissipative forces (4.3), for the more advanced students, with dissipative forces (equation (4.6)(4.9)). For example, we discuss the motion of the parachute and the conditions for reaching the final constant velocity and its dependence on the mass of the suspended object etc. In Figure 4.22 below, an illustration of the exercises given to the students is shown. These are assignments for both the lower and higher level students and reflect their current level of knowledge.

Let us next give an example of the students' work in the gravitational field (Figure 4.23 a and b). This type of tasks often causes problems to the students mainly with the orientation of the coordination axis, as well as the expressions describing the motion, which is axis dependent.

Figure 4.22 _ Exercises for seminar on motion in the gravitation field

Lower level

Exercise 1: The magnet starts to move from the end of the glass tube of the length $L = 1$ m provided with ten thin pick-up coils wound in regular distance of $\Delta y = 10$ cm. Determine:

a) What is the passage time through the whole tube t_p and the time Δt, of passing between the third and fifth pick-up coil,

b) The general expression for the passage time between n^{th} and $(n+1)^{th}$ pick-up coils. The path Δy run by the magnet for the time interval $0{,}015$ after $t = 0{,}2$ s from the beginning of the motion,

c) The time for passing the first 20 cm of the tube. Prove all results using data of remote experiment.

Exercise 2: Using Excell or Origin graphs processing environments plot graph of temporal dependence of position $y(t) = f(t)$ and fit it by the the second order polynom $y(t) = at^2 + bt +c$. Discuss the physical meaning of individual coefficients.

Higher level

Exercise 3: For given length of the tube $L = 1$ m determine:

a) The required length of the evacuated tube in order to achieve the velocity at the end of the tube $v_L = 10$ m.s^{-1},

b) The coefficient of friction force k_1 (eq. 4.6) for the magnet to start to move approximately by constant velocity,

c) The heat developed in b) from the start till this point and from this point by unit length of the tube.

Exercise 4: By means of calculus derive expressions for kinematics quantities describing motion of a body in Earth' gravitation field, both taking into consideration and neglecting dissipative forces, all in approximation of mass point (i.e. neglecting its dimensions).

Use simulation on Free fall: http://jersey.uoregon.edu/AverageVelocity/

a) Exploring the motion of the ball during the free fall time dependence and average velocity

The simulation shows motion of the falling body and helps to solve problems "how long it takes for a body to cover the pass between two points, e.g. the third and fourth meter of its trajectory". Let us choose the positive orientation of y – axis in the direction of the falling body and its beginning at the start of the motion at $y_0 = 0$ m and $t = 0$ s. The time Δt elapsed to cover the path $s = y_2 - y_1$ between two arbitrary points y_1 and y_2 is

$$\Delta t = t_2 - t_1 = \sqrt{\frac{2}{g}} \left(\sqrt{y_2 - y_0} - \sqrt{y_1 - y_0} \right).$$

The simulation helps to prove this expression.

The simulation and calculation:

The table shows the measured and calculated times to cover increasing paths of the ball at different celestial bodies, the Earth (E) – $g_E = 9.81$ ms^{-2}, Moon (MO) – $g_{MO} = 1.62$ ms^{-2} and Mars (MA) – $g_{MA} = 3.75$ ms^{-2}.

Path $s = y_i - y_0$ $i = 1,..., 6$ [m]	Time t to cover the path [s]			
	The Earth		Moon	Mars
	Exp.	Calc.	Exp.	Exp.
1	0.45	0.449	1.11	0.73
2	0.63	0.638	1.57	1.04
3	0.78	0.782	1.92	1.28
4	0.90	0.903	2.22	1.47
5	1.01	1.010	2.48	1.65
6	1.10	0.096	2.72	1.81

Average velocity for free fall from 6 m height:

Height $h = 6$ m	Average velocity v_a [m s^{-1}]		
	The Earth	Moon	Mars
	5.45	2.20	3.31

The presumption $v_E > v_{MA} > v_{MO}$ was verified by both, the measurement and the calculation.

Figure 4.23a _ Exercises for the seminar on motion in the gravitational field

Continuation of the previous panel

Use simulation on free fall: http://jersey.uoregon.edu/AverageVelocity/

b) Finding the landing velocity v_l

Next we want to calculate the landing velocity of the ball falling by free fall from height h = 6 m on all the explored space objects [the Earth (E), Moon (MO) and Mars (MA)]. Make a guess and verify it by means of the simulation.

My guess for landing velocity: $v_{lE} > v_{lMA} > v_{lMO}$

For the landing time t_l we use eq.

$$t = \sqrt{2y/g}.$$

By the derivative with respect of time t is for landing velocity v_l and time t_l :

$$v_l = \sqrt{2hg} \quad and \quad t_l = v_l/g.$$

Experiment and calculation: The results are summarized in the table.

Height h = 6 m	Landing time t_l [s]		
	The Earth	Moon	Mars
Calculation	1.03	2.58	1.73
Simulation	1.05	2.71	1.80
Landing velocity v_l [ms^{-1}]	11.43	4.43	6.67

The presumption $v_E > v_{MA} > v_{MO}$ was verified by both, the measurement and the calculation.

c) Finding the acceleration due to gravity on all the explored space objects.

My guess: $g_E > g_{MA} > g_{MO}$

We use the equation: $g = (2h/t_l^2) \rightarrow g_E > g_{MA} > g_{MO}$

d) Is the gravity on Moon six times weaker than the Earth gravity?

9.92 ms^{-2} / 1.63 ms^{-2} = 6.08 → yes

e) Finding the mass of the Moon and Mars.

In part b) we found the acceleration due to gravity on the Moon (Mars). When the values of the diameters r of these objects are known, we can calculate their mass by means of the equation: distances form the Earth?

$$g = \frac{2h}{t_l^2} \Rightarrow g_Z > g_{MO} > g_{MA} \qquad g = G\frac{mh}{r^2} \Rightarrow m = g\frac{r^2}{G}.$$

Figure 4.23b _ Exercises for the seminar on motion in the gravitational field

C) Laboratory exercises – In the laboratory exercises there is enough time for the comparison of experimental observations and model calculations using the theory delivered in the lectures. A student for example, exerts the measurements using RE "Free fall" using both the air ambient and evacuated tube, then observes, explains and evaluates the differences in both measurements. How demanding the assignment is depends on the level of the student – researcher – e.g.:

"Examine the kinetics of the observed motion, determine deviations from the ideal free fall and discuss the reasons for it".

"Pay attention to the energy of the free fall in an ideal scenario and determine the free fall acceleration with the focus on the differences determined from the local table value."

"Analyse the free fall with the presence of the dissipative forces, determine the free fall local acceleration and find by extrapolation the limiting velocity."

"In the case of the presence of dissipative forces, discuss the energy losses and creation of heat caused".

D) Project – An indispensable part of the subject Physics 0 – Mechanics are the research-based activities in the form of the students' project, where virtual and real remote experiments are taken as experimental instruments and the "output" data is taken as measured and critically compared with the corresponding theoretical models.

During the semester, the students download the projects by means of the LMS MOODLE in prescribed form, continuously corrected by the instructor with respect to misconceptions, errors and improvement of the formal layout. If the student is willing to cooperate, he/she may constantly improve his/her evaluation.

As an example, let us have a look at the selected part of the students' project "Free fall", which is the virtual experiment by W. Christian and J. Belloni [81]. The simulation presents time dependence of the instantaneous position, velocity and acceleration for two different initial conditions, (http://zamestnanci.fai.utb.cz/~schauer/contents/mechanics /one_d_kinematics/ illustration2_6.html, Figure 4.24 simulation 1 – left, simulation 2 – right). The student should analyse and derive the initial conditions for the simulation. The simulation is very similar to our RE "Free fall", and may serve as a basis for discussion about it.

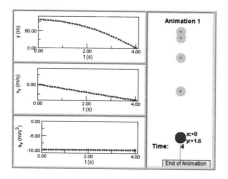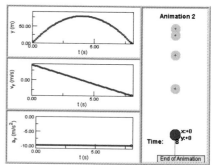

Figure 4.24 _ Simulation of motion in the gravitational field

Examples of students' project work with simulation 1 and 2 is shown in Figure 4.25.

The goal of the project was to explore three cases of the free fall by means of the simulation. I focused on the quantities of the velocity and acceleration found by means of derivatives. In all three cases of motion are different initial conditions.

Table of the calculated data for the simulation 1 in Fig. 4.24 left

t [s]	0	1	2	3	4
y [m]	80	75.1	60.4	35.9	1.6
vy [ms^{-1}]	0	-9.8	-19.6	-29.4	-39.2
a [ms^{-2}]	-9.8	-9.8	-9.8	-9.8	-9.8

The calculated values show the growing decrease of the co-ordinate y with the time, the linear decrease of the negative velocity vy with time, because of the orientation of the y axis and the constant value of the acceleration (a_y = – 9.8 ms^{-2}). The x-axis is not considered as we observe the free fall along y axis only. The differences of the theoretical and "experimental" data are negligible, testifying the correctness of simulation.

Table of the calculated data for the simulation 2 in Fig. 4.24 right:

t [s]	0	1	2	3	4	5	6	7	8
y [m]	0	34.3	58.8	75.5	80	75.5	58.8	34.3	0
v [ms^{-1}]	39.2	29.4	19.6	9.8	0	-9.8	-19.6	-29.4	-39.2
a [ms^{-2}]	-9.8	-9.8	-9.8	-9.8	-9.8	-9.8	-9.8	-9.8	-9.8

The ball in simulation 2 is thrown upwards with the initial velocity v_0 = 39.2 ms^{-1}. The velocity decreases linearly in time till the ball reaches its maximum height, where is the velocity zero. Then the velocity starts to increase linearly in the opposite direction. The ball moves with the constant acceleration a_y = g = $-$ 9.8 ms^{-2}. The maximum height of the ball is y_{max} = 80 m where its velocity is v_y = 0 ms^{-1}. The ball reaches its starting point (where we defined y = yo = 0 m) with the velocity vo but with the opposite direction.

The correctness of the simulation was proved by the calculation as the difference were negligible. The work on the project was the enrichment of the study of Physics and I could also deepen the basic knowledge from Kinematics. The simulation is well designed and it's a great tool to verify to elucidate the kinematic of free fall motion.

Figure 4.25 _ Illustration of the work with simulation 1 and simulation 2

E) Examination – With respect to the research-based way of teaching, even the method of evaluation of acquired students' knowledge and skills should be different from that of the traditional way of teaching. In this respect, the interactive simulations for the examination sand the use of the REs were distinctly advantageous. The teacher may focus on those activities the students has had to master. In any ensuing discussion about the live simulations, the teacher easily recognises the student's understanding of the subject matter and his/her ability to work with the data, it's processing and evaluation. This method of knowledge feedback is very important for the pre-service physics teachers as they are provided with useful methods for their future teacher's praxis.

F) Knowledge self-control –The student has at his/her disposal, as a part of the study materials for the Mechanics course, and other parts of the Bachelor Physics course, the set of test questions for self-control of acquired knowledge delivered by both the MOODLE questioner, or our MODDLE module IPAL for wireless clickers [38].

4.4 Research-based teaching – "Oscillations"

The experimental basis for research-based activities in periodic motion is RE "Simple pendulum". The RE is accessible at the following URL address: http://remotelab5.truni.sk.

After choosing and giving an initial deflection by the robotic trolley, the researcher (Figure 4.26 left) starts the experiment. The client may then observe the running of the experiment by the online camera (Figure 4.26 middle). Each swing is picked up by the optical gate, depicted in the graph by the red vertical lines. The uniqueness of the experiment is in the measurement of the instantaneous deflection $\phi = \phi(t)$ by the reconstruction using the algorithm given in [88], from the data of two dynamometers D_1 a D_2

$$\varphi = \left[\frac{\delta}{2} - \arcsin\left(\frac{F_2 \sin(\pi - \delta)}{F_1 + F_2} \right) \right], \tag{4.14}$$

where δ is the angle between the forces of the pull (tension) F_1 a F_2 (Figure 4.27 right).

In Table 4.3 the time dependences of the instantaneous deflection $\phi(t)$ and the number of exerted swings are summarised.

Let us now show how we can express all the quantities characterising a simple pendulum using the instantaneous deflection angle φ (Figure 4.27).

For the case of small deflections $\varphi < 5°$ the motion is possible to describe by

$$\varphi = \varphi_0 \sin(\Omega t) \tag{4.15}$$

Figure 4.26 _ RE "Simple pendulum"– detail of the experiment: the unit for re giving the pendulum the initial deflection of the pre-selected value withthe step motor controlled motion (1), position sensing resistor (2) and electromagnet (3) – (left), graphical time representation of the instantaneous deflection $\varphi(t)$ – (middle), board with two dynamometers (D1 and D2) for reconstruction of the instantaneous deflection $\varphi(t)$ of the pendulum (right)

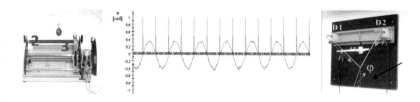

Table 4.3 _ Example of the experimental data from RE "Simple pendulum"

Time t (s)	Deflection φ (rad)	Number of swings	Time t (s)	Deflection φ (rad)	Number of swings
0.0078	-0.2148	1	0.1070	-0.3398	1
0.0273	-0.4375	1	0.2266	-0.3281	1
0.0469	-0.4062	1	0.2500	-0.3477	1
0.0664	-0.4141	1	0.2695	-0.3945	1
0.0898	-0.3047	1	0.2891	-0.3398	1
0.1094	-0.3594	1	0.3086	-0.3516	1
0.1289	-0.4336	1	0.3281	-0.2695	1
0.1484	-0.4023	1	0.3477	-0.3047	1
0.1680	-0.4102	1	0.3672	-0.3008	1
0.1875	-0.3125	1	0.3867	-0.3281	1

where φ_0 is the amplitude, $\Omega = 2\pi f$ is the angular velocity and f is the frequency

$$\Omega = \sqrt{\frac{g}{r}}, \qquad (4.16)$$

where r is the length of the pendulum pull and g is the acceleration of gravity.

How can the teacher explore the above described experiment in the teaching of Mechanics? Let us show successively using the RE in research-based method by means of INTe-L strategy in all forms of the teaching of the unit Mechanics.

A) Lecture – In the lecture we start with RE for the demonstration, discussion and also its' technical realisation. Then follows the individual topics of kinematics, dynamics and energy with the work of forces describing the phenomenon of oscillations.

Kinematics of the curvilinear motion

The aim of the kinematic approach is to explain the basic kinematics quantities characterizing curvilinear motion. We make use of the motion of a simple pendulum, depicted in Figure 4.27. The pendulum moves with the harmonic motion described by the equation (4.15). (Notice the positive and negative sign of the deflection $\varphi >$ or < 0).

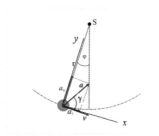

Figure 4.27 _ To the explanation of tangential a_t and normal acceleration a_n of curvilinear motion

We may split the acceleration a into two components, tangential a_t and normal a_n. The basic characteristic of curvilinear motion, its normal acceleration a_n, differs from zero. It is caused by the fact that the direction of the unit vector of instantaneous velocity τ, given by the tangent to the trajectory ($v = v\tau$, where v is velocity value), is changing with the time

$$a_n = \frac{d\left(\dfrac{v}{v}\right)}{dt}$$

(4.17)

The value of the normal acceleration is then

$$a_n = \frac{v^2}{r},$$

(4.18)

with its direction normal to the tangent and where r is the length of the suspension of the simple pendulum.

The tangential acceleration, expressing the change of the velocity value, is defined by the expression

$$a_t = \frac{dv}{dt},$$

(4.19)

and its direction is given by the tangent to the trajectory. The angle γ between tangential and normal accelerations (Figure 4.27), is

$$\mathrm{tg}\,\gamma = \frac{a_n}{a_t}.$$

(4.20)

Based on the above expressions we may determine those quantities for simple pendulum oscillatory movement using the instantaneous deflection (4.15) and the expression for the angular velocity ω

$$\omega = \left(\frac{d\varphi}{dt}\right).$$

(4.21)

The normal acceleration then is

$$a_n = r\omega^2 = r\left(\frac{d\varphi}{dt}\right)^2 = r\left(\varphi_0\Omega\cos\Omega t\right)^2.$$

(4.22)

For the tangential acceleration, we use the expression for the angular acceleration α

$$\alpha = \frac{d^2\varphi}{dt^2}, \Rightarrow a_t = r\alpha = r\left(\frac{d^2\varphi}{dt^2}\right) = -r\left(\varphi_0\Omega^2\sin(\Omega t)\right).$$

(4.23)

The time dependence of the tangential and normal accelerations of the simple pendulum is in Figure 4.28.

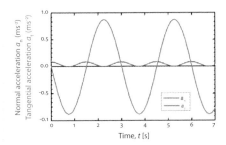

Figure 4.28 _ Time dependence of the tangential and normal accelerations of the simple pendulum (r = 2 m, φ_0 = 0.1 rad, t = 3 s, m = 0.1 kg) determined by equations (4.19) and (4.22)

At the end of the topic "Kinematics of the curvilinear motion" the teacher may ask several check and motivation questions, both as examples or as a problem-solving exercise. Let us illustrate several of them below:

– *Model (using e.g. graphic environment WZGrapher) and discuss time dependence of the instantaneous deflection, velocity and acceleration of the simple pendulum.*
– *A simple pendulum with the length r = 2 m carrying the weight of mass m = 1.5 kg moves with the maximum deflection angle ϕ_0 = 5⁰ and the angular velocity of the oscillatory motion Ω = 3 s⁻¹. Determine the necessary*

minimum strength of the suspension and the point, where the pull will be at maximum.
- *Determine the points of the maximum and minimum velocity and its value.*
- *What can we determine from the graph of the time dependence of the acceleration components? What does it mean that tangential acceleration has both positive and negative signs while normal acceleration exerts only positive values?*
- *Determine the points with maximum and minimum a_t and a_n and their values.*
- *Describe the changes of amplitudes of both the tangential a_t and normal acceleration a_n, with changing the frequency of the pendulum oscillations.*

Dynamics of the curvilinear motion

We are often faced with the following question from the student: "What is behind the fact that the simple pendulum is moving with harmonic motion?" The answer may be given, based on the 2nd Newton equation of motion: *The body acceleration a is proportional to the sum of all functioning forces and inversely proportion to its mass m. For the simple pendulum it is* (see Figure 4.29)

$$a = \frac{\Sigma F}{m} = \frac{F_G + F_T}{m}, \tag{4.24}$$

where F_G is the weight and F_T is the force of pull.

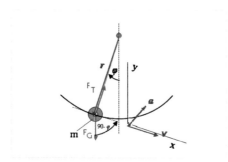

Figure 4.29 _ Schematic illustration of forces acting on a Simple pendulum

For the simple pendulum in Figure 4.29 it is possible to split the vector equation 4.24 into two scalar equations for x and y axis:

$$\text{for } x\text{: } (\varphi < 0)\text{: } F_G \sin \varphi = -ma_t, \tag{4.25}$$

$$\text{for } y\text{: } F_T - F_G \cos \varphi = -ma_n. \tag{4.26}$$

Equation 4.25 is in fact the differential equation of motion

$$\frac{d^2\varphi}{dt^2} = -\frac{g}{r}\sin\varphi,$$

which for small deflections ($\varphi < 0{,}1$ rad, or $\varphi < 5°$ is approximately $\sin \varphi \cong \varphi$) is

$$\frac{d^2\varphi}{dt^2} = -\frac{g}{r}\varphi. \tag{4.27}$$

Its solution is the harmonic function (Equation 4.14)

$$\varphi = \varphi_0 \sin(\Omega t),$$

where the quantity of the angular velocity Ω is determined by the expression 4.16

$$\Omega = 2\pi f = \sqrt{\frac{g}{r}}.$$

Equation 4.22 gives for the value of the pull force on the moving body

$$F_T = mg \cos\left[\varphi_0 \sin(\Omega t)\right] + mr\left[\varphi_0 \Omega \cos(\Omega t)\right]^2. \tag{4.28}$$

Its time dependence is shown in Figure 4.30. It is interesting to return to the principle of the measurements of the instantaneous deflection angle φ determined from the measurements of two dynamometers F_1 and F_2

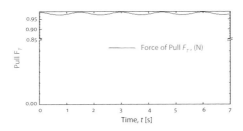

Figure 4.30 _ Time dependence of the pull force of moving pendulum

(Equation 4.14 and Figure 4.26 right) $F_{T}' = F_{1} + F_{2}$. From the law of action and reaction it is for the values of both forces $F_{T}' = F_{T}$.

Energetic aspect of the curvilinear motion

We can express the kinetic energy of the pendulum by the angular velocity of the motion ω

$$E_k = \frac{1}{2}mv^2 = \frac{1}{2}m(r\omega)^2 = \frac{1}{2}m\left[r\varphi_o\Omega\cos(\Omega t)\right]^2 \tag{4.29}$$

and the gravitational potential energy E_p of the weight using the instantaneous position h above the arbitrary chosen reference level (here it is the distance to the lowest point of the pendulum $h = r[1 - \cos\varphi(t)]$) in Figure 4.31, it is

$$E_p = mgh = mgr\left[1 - \cos(\varphi(t))\right] = mgr\{1 - \cos\left[\varphi_0\sin(\Omega t)\right]\}. \tag{4.30}$$

For the motion of the pendulum we can verify the validity of the energy conservation law expressed by the expression

$$E_{k1} + E_{p1} = E_{k2} + E_{p2}, \tag{4.31}$$

where E_{k1}, E_{k2} are the kinetic energies in points 1 and 2, and E_{p1}, E_{p2} are the corresponding potential gravitation energies of the weight with respect to the lowest point of the pendulum taken as reference energy. Time dependence of the pendulum energies are in Figure 4.32.

B) Seminar – The students in seminars verify the knowledge gained from the lessons and how this knowledge strengthens them by the application

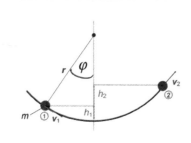

Figure 4.31 _ The energy conservation law of the simple pendulum

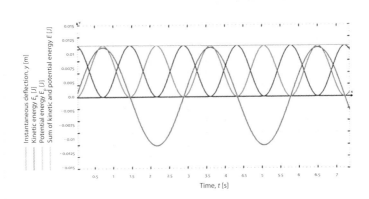

Figure 4.32 _ Time dependence of kinetic energy E_k (blue curve), potential gravitational energy E_p (green curve) and total mechanical energy E_c (orange straight line), instantaneous deflection y (red curve) of the simple pendulum with real parameters of RE r = 2 m, φ_0 = 0.1 rad, T = 3 s, m = 0.1 kg

to the real world problems around us. A lot of interesting examples exist. Besides real REs, virtual experiments from W. Fendt's webpage (http://www.walter-fendt.de/ph14cz /pendulum_cz.htm) are used, together with PhET team (http://phet.colorado.edu), as well as from the monograph by W. Christian and M. Belloni [6] available also at the TBU repository http://zamestnanci.fai.utb.cz/~schauer/contents/oscillations_waves/periodic_motion/ex16_4.html.

These simulations or REs serve as the starting point for the calculation of simple or more advanced exercises. In Figure 4.33 and Figure 4.34 the examples of exercises on the oscillations of simple pendulum are shown.

C) Laboratory exercise – The acceleration of the gravity g in the given locality with the maximum precision is another possibility of this RE use in combination with the RE already presented in the study of g by means of free fall in the evacuated tube. An example of experimental data processing and an evaluation from the student's laboratory work is shown in Figure 4.35.

D) Project work – The student has to master fitting of the data by the theoretical curves and data presentation in a graphical environment. The RE enables the student to study the influence of the dissipative media and discuss its impact on energy conservation. An example of the damping and the calculation of the damping coefficient *b* and the resulting acceleration of gravity *g* is shown in Figure 4.36 A and Figure 4.36 B. As a result of the small damping exerted by the simple pendulum, this enables measuring over a prolonged period of time and hence more accurate results.

Let us select a comment from the conclusions of the student's report:

„This experiment was functioning without any major problems. Data transfer from the measure server was also without problems and the transferred data was easy to analyse. We only selected the data after reaching equilibrated motion as only this represents a true harmonic motion. Thanks to the negligibly small irregularity in the oscillatory motion, the resulting quantities were also relatively precise. That is why I evaluate the laboratory exercise as a success.”

Figure 4.33 _ Example of exercise for the seminar on a simple pendulum

Problem 1:

Prove the functionality and mathematical validity of the simulation "Simple pendulum" (http://www.walter-fendt.de/ph6en/pendulum_en.htm) and calculate the length of the pendulum L when it makes 5 oscillations in 22.5 s.

Solution:
As we know, the frequency is given $f = 5/22.5$ s = 0.222 Hz, so the period of the oscillations is $T = 1/f = 4.5$ s.
The length of the pendulum is

$$T = 2\pi\sqrt{\frac{L}{g}}$$

$$L = \frac{T^2 g}{4\pi^2} = \frac{4,5^2 s^2 \cdot 9,81\,ms^2}{4.3,14^2} = 5\,m.$$

The calculation is proved by the simulation in the picture:

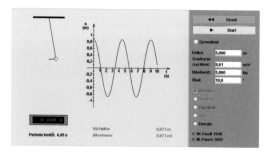

Problem 2:

A body of the mass $m = 5\,kg$ oscillates with the period $T = 4.44$ s. The total energy of the body is 0.0378 J. Find the equation for the deflection of the body and the maximum value of the force acting on the body in the approximation of a point mass.

Solution:
If we know the instantaneous deflection y, the total energy of the oscillator E and the spring constant k, we can write:

$$y = y_m \sin \omega t, \; E = \frac{1}{2}ky_m^2, \; k = \omega^2 m.$$

From these equations then

$$E = \frac{1}{2}ky_m^2 = \frac{1}{2}m\omega^2 y_m^2 = \frac{1}{2}m\frac{4\pi^2}{T^2}y_{max}^2$$

$$\omega = \frac{2\pi}{T} = \frac{2.3.14}{4.44\,s} = 1.414\,s^{-1}$$

$$y_{max} = \sqrt{\frac{2T^2 E}{4\pi^2 m}} = \sqrt{\frac{2.(4.44\,s)^2.0.0378J}{4.(3.14)^2.5\,kg}} = \sqrt{0.00757}\,cm = 0.087\,m$$

$$k = \omega^2 m = (1.414\,s^{-1})^2.5\,kg = 10\,kgs^{-2}.$$

The equation of the deflection of the body is: $y = 0.087 \sin(1.414\ t)$. For the value of the maximum force, acting on the body, we can write:

$$F_{max} = ky_{max} = 10\,kgs^{-2}.0.087\,m = 0.87\,N.$$

This value may be proved by the following simulation:

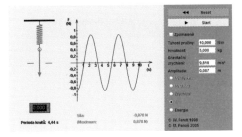

Problem 3:

A simple pendulum of the length $L = 2$ m carrying a body of the mass $m = 1.5$ kg oscillates with the maximum amplitude of the angular displacement $\varphi_o = 5°$ and with the angular frequency $\Omega = 3\,s^{-1}$. Find the minimum strength of the hinge and the place in which the hinge is stressed most. Use WZGrapher environment to model the situation, and discuss the force of tension F_T.

Figure 4.34 _ Example of exercise for the seminar using simulation

Experimental data graph processing

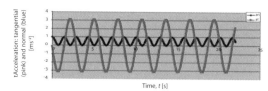

Graph 1: Time dependence of the tangential and normal acceleration

Graph 2: Time dependence of the tension force in suspension

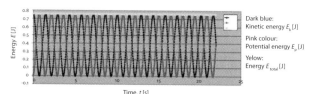

Graph 3: Time dependence of the kinetic, potential and total energy

4. Calculation

L = 2.07 m

T = 2.88828 s

$$T = 2\pi\sqrt{\frac{L}{g}} \Rightarrow g = 4\pi^2\frac{L}{T^2}$$

$$\frac{\Delta g}{g} = \frac{\Delta L}{L} + \frac{\Delta T}{T} \rightarrow$$

g = (9.796 ± 0.199) ms^{-2}

Figure 4.35 _ Example of data evaluation from the laboratory exercise

Assignment

— Make an orientation measurement,
— Find the damping coefficient of the simple pendulum.

Measurement

For the sample measurement we chose the measurement for 1200 s corresponding to the total number $N = 415.5$ periods and made the graph.

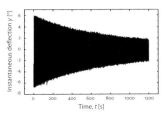 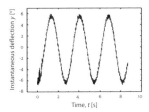

Fig. 19 Graph of the sample measurement of RE "Simple pendulum"

The pendulum oscillates with a damping, so we decided to calculate the damping coefficient b by fitting the envelope of the oscillations by the exponential function and gained $b = 0.0012 \ s^{-1}$.

— Find the acceleration due to the gravity,
— Plot the graphs of the kinetic, potential and total energy.

Figure 4.36a _ Example of exercise for the seminar using simulation

Assignment (continued)

The table of the measured and calculated values of the acceleration due to the gravity

No.	Number of oscillations N	Measurement duration t (s)	Period of oscillation T (s)	Gravity acceleration g (ms^2)	Abs. unc. Δg (ms^2)
1	800	2311.7	2.8898	9.7864	0.00004
2	750	2167.3	2.8897	9.7860	0.00000
3	700	2023.0	2.8900	9.7843	0.00017
4	650	1878.5	2.8900	9.7844	0.00016
5	650	1878.4	2.8898	9.7858	0.00002
6	650	1878.0	2.8893	9.7893	0.00034
7	650	1878.3	2.8897	9.7864	0.00004
Average	–	–	2.8897	9.7860	0.00011

The table shows the results of the measurements. Based on the table data we calculated the most probable acceleration due to the gravity with three decimal places of uncertainty

$g = (9.786 \pm 0.002)$ ms^{-2}.

The resulting equation of the fitting function for oscillation of the damped simple pendulum is:

$y = 0.1028e^{-0.00117t} \sin (2.1744t + 1.5708)$.

Prolonged measurements enable us to observe the decline of the total energy even with such a small damping (see the graph below).

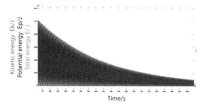

Time dependence of kinetic, potential and total energy

Figure 4.36b _ Example of data processing of the damped simple pendulum for coefficient of damping in project work (continuation of the previous panel)

4.5 Research-based teaching – "Dissipative processes"

In this part of Chapter 4, we want to show the relation of the macroscopic energy and energy of the microphysical dissipative processes using RE in the "Joule experiment".

Based on the experiment of the transformation of mechanical energy into heat energy, J. P. Joule has formulated the energy conservation law. His experiment, which dates back to 1840, is designed as a system of rotating paddles in a thermally isolated vessel filled with water, arranged so that they can revolve (Figure 4.37). The rotation is due to the system of two weights moving downwards, so that the transformation of the potential energy of the gravity of the weights to the kinetic energy of the rotating paddles takes place with a successive change to the heat.

The whole phenomenon is described by a general equation

$$\Delta E_p + \Delta E_k + Q = 0 \, , \tag{4.32}$$

where ΔE_p is the increase of potential energy of gravity of weights (negative) and ΔE_k is the increase of the kinetic energy of the rotating parts of the apparatus and Q is the heat evolved during the experiment. Results of the experiment led Joule to the conclusion about heat as another form of energy. Later it was decisively proved that heat is the total kinetic

Figure 4.37 _ Drawing of the Joule apparatus
(http://atropos.as.arizona.edu/aiz/teaching/nats102/mario/matterenergy.html)

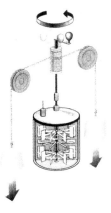

energy of the heat motion of all particles of the system. Today it is commonly expressed as

$$Q = mc\Delta t, \tag{4.33}$$

where m is the mass and c is the specific heat of water and Δt is the increase of its temperature. Today we know that an energy of 4,186 J supplied to 1 kg of water results in a temperature increase of 1 °C, so water specific heat is c = 4186 $J.kg^{-1}K^{-1}$.

RE was set up from the components of fy PHYWE and for the data collection system ISES (Figure 4.38). The basic goal of the Joule RE is to find the relation between the work by dissipative forces of the friction and the evolved heat.

The principle is obvious from the simulation in Figure 4.39, (http://homepages.ius.edu/kforinas/contents/thermo/heat/prob19_2.html), where a blue transportation belt is in contact through friction with the stationary red body. The heat generated by this process is manifested through the temperature increase. From the exerted work known it is possible, using the known mass of the body, to determine the specific heat of the red body neglecting the heat losses to the environment.

In our RE, the driving motor (6) in Figure 4.38 drives the rotating cylinder made of various materials (brass, aluminium and other metals). The

Figure 4.38 _ Total view of the Joule experiment (left); detailed arrangement (right)

1 – Hands on experiment, 2 – ISES interface, 3 – PC (ISES)

1 – Dynamometer, 2 – Voltage source, 3 – Roller,
4 – Optical gate, 5 – Thermometer, 6 – Driving motor (12 V; 2.2 A),
7 – Connecting wires, 8 – Weight, 9 – Plastic tape

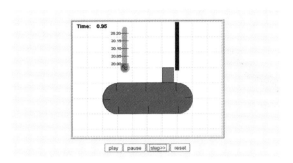

Figure 4.39 _ Simulation "Joule experiment"

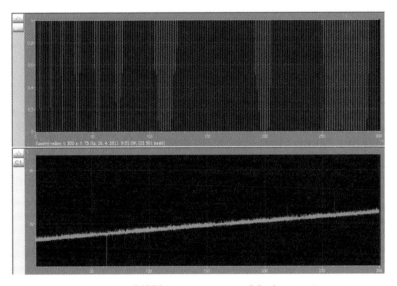

Figure 4.40 _ Example of ISES representation of Joule experiment measurement – number of revolutions (up); time dependence of the temperature (down)

friction force is equal to the difference between the force F_2 measured by the force meter and the weight of the used weight F_1.

$$F_T = F_2 - F_1 .$$

The work of friction forces is

$$W = 2 \pi r n (F_2 - F_1),$$ (4.34)

where r is the cylinder radius, and n is the number of revolutions. The work

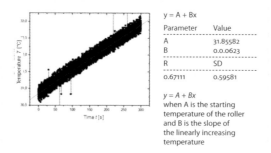

y = A + Bx

Parameter	Value
A	31.85582
B	0.0.0623
R	SD
0.67111	0.59581

y = A + Bx
when A is the starting
temperature of the roller
and B is the slope of
the linearly increasing
temperature

Figure 4.41 _ Example of exercise for the seminar using simulation

of the friction force and thus the generated heat may be adjusted by the elapsed path $2\pi rn$ and by the mass of the weight used.

Examples of the measured data in graphical form for the time t = 300 s and the number of revolutions n = 170 measured by ISES are shown in Figure 4.40 and Figure 4.41. The first panel (upper part of Figure 4.40) depicts the number of revolutions measured by the optical gate and the corresponding increase of the temperature T of the cylinder with the time t depicted in the lower panel.

In Figure 4.41 we present an illustrative example with its parameters and results gained by means of the equations (4.32) and (4.33) for the cylinder made of aluminium, weight of the mass was 1.2 kg, the time of measurement t = 500 s, the number of revolutions n = 250, the frequency f = 2 s^{-1}. The mass of the aluminium cylinder m_{Al} = 0.43 kg, its specific heat c_{Al} = 896 J.kg^{-1}.K^{-1}, the cylinder radius r = 0.023 m, the starting temperature T_1 = 37.3 °C, the final temperature T_2 = 40.6 °C, the force F_2 = 50 N, measured by the force meter and F_1= 12 N the force of the used weight. The measured evolved heat was

$$Q = 0.43 \text{ kg} .896 \text{ J} / \text{kg K}. 3.31 \text{ °C}$$
$$Q = 1275.28 \text{ J},$$

and the exerted work was

$$W = 2 . 0.0225 \text{ m}.250 \text{ Hz}.38 \text{ N}$$
$$W = 1372.18 \text{ J}.$$

We can see the relative difference was about 8 % and the agreement is quite satisfactory, caused by not taking into consideration the losses of heat to the environment.

A) Lecture – In physics, going from any arbitrary idealised phenomenon to a real one we have to stress several important circumstances concerning the transformation of the energy from one form to another. In the idealised case, the phenomenon takes place without any energy losses that are neglected. In real world phenomena losses always exist that decrease the energy transformation, sometimes substantially. We may also mention that we can obtain a lower electric energy from a rechargeable battery due to the evolved heat generated during its discharge. Next examples may be the heating of the computer, the hot motor in the refrigerator or the limited efficiency of solar cells. This is a reason why omnipotent dissipative losses in phenomena are important for study of real world situations. The Joule experiment has played an important role in this direction, as it has showed where there are reductions in the energy lost – for the heating of bodies of the phenomenon itself and their surroundings. The Joule experiment can help explain these dissipative processes and the teacher's role is to generalize them for the subject matter being taught.

B) Seminar – An example of students' work is shown in Figure 4.42. The teacher, together with the students, may discuss the efficiency of energy transformation (e.g. electrical to mechanical, electrical to light, electrical to the heat taken away....) or the topic *"Why are electrical appliances distributed in several efficiency categories A – F, similar to the marks given at the examination?"*

The next interesting topic for students may be the branch of photovoltaics. The teacher may pose questions such as:

- *Give example of semi-conductors used in photovoltaics and compare the efficiency of transformation of the sun radiation to electrical energy for semiconductors: Si, GaAs and Si:H (silicon, gallium arsenide and amorphous hydrogenated silicon).*

- *Give examples of phenomena with extremely high (approaching 100 %) and very low efficiency;*

- *What may be the efficiency of biological processes?*

- *What is the relation between energy intake and motion or specifically intake of foodstuff and energy output during physical exercise; As an example we can present the simulation* (Figure 4.43).

Problem

http://pages.iu.edu/~kforinas/contents/thermo/heat/prob19_2.html

A block with mass m = 4.3 kg made of material with specific heat capacity c = 234 J.kg^{-1}.K^{-1} has the starting temperature T = 20.00 °C. We observed, that by moving the belt the temperature increased by DT when running the path l. Find the coefficient of the friction between the block and the belt. See picture of the simulation down.

Verify the validity of the simulation by measuring the change of the temperature of the body when you know the value of the coefficient of the friction f = 1.19.

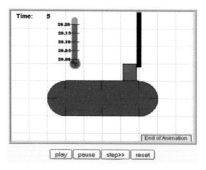

Solution:

For calculation we use the following values of the quantities: c = 234 J.kg^{-1}.K^{-1}, l = 1 m, m = 4.3 kg, g = 9.81 ms^{-2}. We "measured", using the simulation, the increase DT= 0.05 K for the path l = 1 m. The coefficient of the friction is given as the quotient of the value of the friction force F_t and normal force acting on the body from the belt F_n = mg.

Work of the friction force, W, is equal to the heat Q, increasing the body temperature, l is path covered by the belt. So it is

$$mc\Delta T = F_t l \quad \rightarrow mc\Delta T = fF_n l \rightarrow f = \frac{c\Delta T}{gl}.$$

Inserting the values of the quantities given above we have for the friction coefficient value

f = 1.19.

Figure 4.42 _ An example of students' seminar work

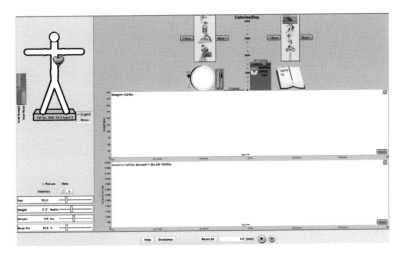

Figure 4.43 _ Simulation "Eating and exercise"
http://phet.colorado.edu/en/simulation/eating-and-exercise

C) Laboratory exercise / project – Similarly, as we have shown in the previous sections, this experiment can be used with multiple assignments, depending on the creativity of the teacher and of the student. The simplest assignment is the measurement of three cylinders with different specific heat capacities. This assignment was part of the project "*Dissipative forces and the heat*" for a pre-service student of Physics of the Faculty of Education in Trnava for the Competition of the best students' work in 2010 and 2011and also for the Czecho-Slovak Competition organised by the Faculty of Natural Sciences of Pavol Jozef Šafárik University in Košice in the same year (Figure 4.44). The participants of the Competition of the best students' work in Trnava 2011 are shown in Figure 4.45.

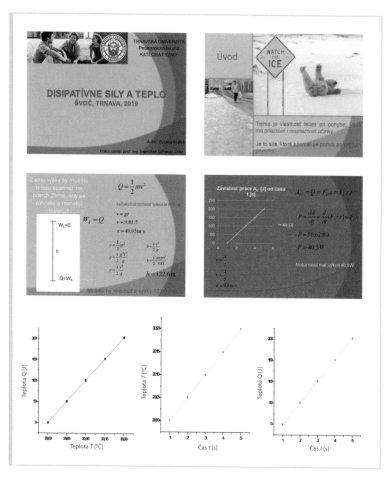

Figure 4.44 _ An example of students' work for the 5th Student's competition, Faculty of Education Trnava University 2011, and Czech-Slovak student's competition, Košice, 2011

Figure 4.45 _ Participants of the 5th Competition of Department of Physics, Trnava University in 2011 (on the whiteboard – online participant on Erazmus stay in Portugal), (photo M. Ožvoldová, 2011)

4.6 Research-based teaching – "Electric circuits"

The experimental basis for the research-based elucidation of the proper-ties of passive electric circuits are phenomena in different combinations of their constituent parts – resistors (R), inductors (L) and capacitors (C). Several REs have been built at the Department of Physics, Faculty of Educa-tion, Trnava University moving in this direction and these are described in the Appendix. For the purpose of this module let us discuss the free acces-sible RE "Transient Phenomena in electric oscillators" (http://remotelab6.truni.sk).

In this RE, behaviour of RLC circuits is examined in the time domain as a response to the applied voltage step. The circuit used in Figure 4.46 contains the source of DC voltage and relay and measuring modules ISES (V-meter and A-meter) for pick-up of the time responses as well as constitu-ent RLC elements of the circuit. The goal of the experiment is to determine the individual parameters of the RLC circuit from the time responses for varying damping due to the variable resistors R_{1D} and R_{2D}.

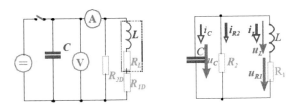

Figure 4.46 _ Parallel RLC circuit with artificial damping due to R_{1D} and R_{2D} (left) and its equivalent circuit (right)

The circuit in Figure 4.46 may be described by the system of Kirchhoff equations I and II

$$0 = i_C + i_L + i_{R2} \tag{4.35}$$

$$u_L + u_R = u_C = u_{R2} = u \, , \tag{4.36}$$

where

$$u_L = L \frac{di_L}{dt} \, , \ \frac{du_L}{dt} = L \frac{d^2 i_L}{dt^2} \, ,$$

$$u_{R1} = R_1 i_{R1}, u_{R2} = R_2 i_{R2}$$

$$i_C = C \frac{du_C}{dt},$$

which gives the final second order differential equation for u

$$\frac{d^2 u}{dt^2} + 2b \frac{du}{dt} + \omega_0^2 u = 0,$$

with solution for $\omega_0 > b$

$$u(t) = u(0) e^{-bt} \sin(\omega_1 t + \varphi). \tag{4.37}$$

In the equation (4.37) $u(0)$ is the step voltage, ω_1 the angular frequency of the RLC circuit,

$$\omega_1^2 = \omega_0^2 - b^2,$$

and the natural frequency ω_0

$$\omega_0^2 = \frac{1}{LC}(1 + \frac{R_1}{R_2}).$$

where $R_1 = R_{1D} + R_L$, $R_2 = R_{2D}$ and φ is the initial phase. The damping coefficient b, $[b] = s^{-1}$, can be thus expressed by the values of all constituent parts of the RLC circuit

$$2b = \frac{1}{R_2 C} + \frac{R_1}{L}. \tag{4.38}$$

The damping coefficient b may be determined by the quotient of two successive amplitudes (at t and $t + T$)

$$\frac{u(t)}{u(t+T)} = e^{-bT}. \tag{4.39}$$

The procedure of the experiment rests in the change of the coefficient of damping b by varying the resistors R_{1D} and R_{2D}. From these dependences we may determine all the values of constituent elements of RLC circuit in question.

The example of the measured response by ISES system is shown in Figure 4.47.

A) Lecture – At the beginning of the lecture the teacher shows transient measurements using RE and points to the circuit, it's lack of dissipation and its behaviour using the conservation of energy law

$E_C + E_L = $ const.,

and compares it with a real circuit with the energy dissipation represented by heat generation Q

$E_C + E_L + Q = $ const.,

where

$$Q(t) = E_0 \left(1 - e^{-bt} \right)$$

is the increase of the heat evolved in the circuit with time, E_0 is the initial total energy in the RLC circuit.

This corresponds to the decrease of the amplitude of the measured voltage

$$u_0(t) = u_{0(t=0\,\text{s})} e^{-bt}.$$

The time dependence of the decrease of the voltage amplitude in a real circuit with damping coefficient b is shown in Figure 4.48.

A sample discussion question may be for example "*How does the evolved heat depend on the circuit parameters RLC found from the damping of time responses?*" For the more advanced students the teacher may pose a

Figure 4.47 _ Damping coefficient measurement by ises in RLC circuit with resistor $R_1 = 10\ \Omega$

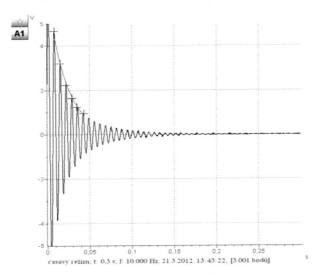

časový režim: t: 0,3 s; f: 10 000 Hz; 21.3.2012 13:43:22; [3 001 bodů]

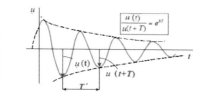

Figure 4.48 _ Oscilations of damped RLC parallel circuit

question how the results of the measurements correlate in the time domain (i.e. the measurements of the response to the voltage step perturbation) and in the frequency domain (i.e. the measurement of the amplitude and initial phase on the frequency).

B) Seminar – As the subject matter for the seminary, teachers may choose the standard exercises for circuits, e.g. from the LMS MOODLE accompanying material from the course Physics I (Figure 4.49) as well as from The Multimedia university physics textbook 2 [90], (Figure 4.50). A more ad-

Figure 4.49 _ An example of the seminar in MOODLE for course Physics I

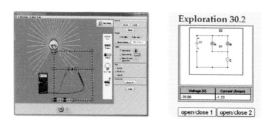

1. The current in a cylindrical copper wire of the diameter $d = 1$ cm is $I = 200$ A. Calculate:

    a) Current density,
    b) How much electric charge Q flows through the wire crossection during the time $t = 20$ s.

2. How much electric charge flows through the wire if:

    a) The electric current linearly increases from zero to 3 A during the time period of $\Delta t = 10$ s,
    b) The electric current decreases from the initial value of 18 A to zero so, that each 0.01 s the current decreases to its half.

3. Solve the problem from applet Pr 30.2 [6] and verify with the measurements; the same with the applet Pr 30.3 available at http://webphysics.davidson.edu/Applets/Applets.html

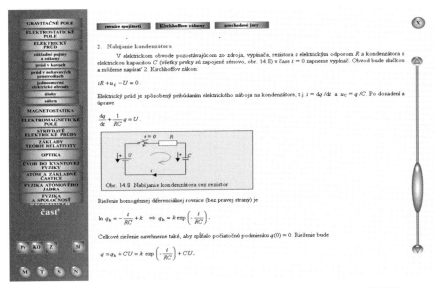

Figure 4.50 _ Example from Slovak "Multimedia University Physics Textbook" [90]

vanced example for the seminar may be the modelling based on the differential equation for RLC circuit stemming from the II. Kirchhoffs'Law, Equation 4.36, using WZ Grapher calculation environment for proving important dependences $\omega_1 = f(b)$ and $b = f(R, L, C)$. A discussion on the comparison of electric and mechanic oscillations would also be a contributing factor.

C) Laboratory exercise – An illustration of the student experimental activity in the laboratory exercise is shown in Figure 4.51 and Figure 4.52, where the dependence of the damping coefficient b on the resistances $R_{1D,}$ resp. R_{2D} are presented in both graphical and tabular form.

The damping coefficient b dependence on the resistances R_{1D} (left) and R_{2D} (right) fitted by Origin software.

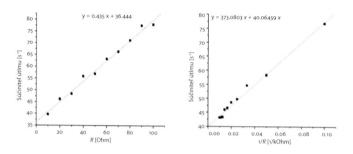

Table: The damping coefficient b as a function of the resistance R_{1D} and R_{2D}

R_{1D} (Ω)	b (s⁻¹)	R_{2D} $(k\Omega)$	b (s⁻¹)	$1/R_{2D}$ $(k\Omega^{-1})$
10	39. 70	10	76.83	0.100
20	46.16	20	58.34	0.050
30	48.40	30	54.63	0.033
40	55.83	40	49.75	0.025
50	56.89	50	48.52	0.020
60	63.18	60	46.49	0.016
70	66.35	70	45.92	0.014
80	71.10	80	43.29	0.013
90	77.44	90	43.14	0.011
100	77.94	100	43.02	0.010

From the measured data follows that the coefficient of damping is realy nearly linearly dependent on both resistances R_{1D} and R_{2D}. For output in graphical form I used graphical environment Origin I have obtained for the straight line fitting

$$y = (36 \pm 1) + (0.43 \pm 0.02)x.$$

For the inductance we than obtain

$$L = (0.86 \pm 0.04) \, H.$$

Figure 4.51 _ Graph and table representation of the dependence of the damping coefficient b as a function of dumping resistances $b = f(R_{1D})$ (left) and b = f(R_{2D}) (right)

Continuation of the previous panel

The point of intersection of vertical axis is $q = (36 \pm 1)$ is

$$\frac{1}{2L} R_L = q$$

and the internal resistance of the inductor than is

$$R_L = 2Lq = (62 \pm 3) \ \Omega.$$

The next task is to find the natural angular frequency in parallel RLC circuit:

$$\omega_0 = 2 \ \pi \ f_0 = 1111.7 \ s^{-1}.$$

By the evaluation of both linearized measurements and using the equation (4.37) we finally obtained the values of individual components:

$$C = 0.99 \ \mu F, \ L = 0.86 \ H, \ R_L = 62 \ \Omega.$$

Figure 4.52 _ An illustration of the student's laboratory calculations of the damping coefficient b (continuation of the previous panel)

4.7 Research-based teaching – "Electromagnetic induction"

The phenomena of electromagnetic induction and the corresponding Faradays' law are the most important parts of physics for their technical applications. Their study is accessible via two REs. One of them is described in the unit 4.3 with RE "Free fall" (http://remotelab4.truni.sk/ in the part devoted to the electromagnetic induction http://remotelab4.truni.sk/faraday.html), so the reader may find corresponding details there. In short, two identical glass tubes (one may be evacuated) are provided with a series of connected pick-up coils with the permanent falling magnet raised to its starting position by an at-home designed magnetic vessel. In this arrangement, the generated voltage on time signal corresponding to the passage of all coils is shown in Figure 4.53 (right) and a live stream of the experiment with controls (left).

The second RE on Faradays' law is the experiment built at the Faculty of Mathematics and Physics at Charles University in Prague (http://kdt-20.karlov.mff.cuni.cz/ovladani_2_en.html) "Electromagnetic induction". The electromagnetic induction there is due to the rotating motion of the coil in

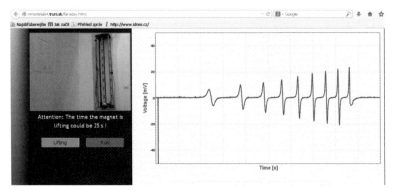

Figure 4.53 _ WWW page RE "Faradays' law"; live stream of the experiment with controls (left) and generated voltage on time (right)

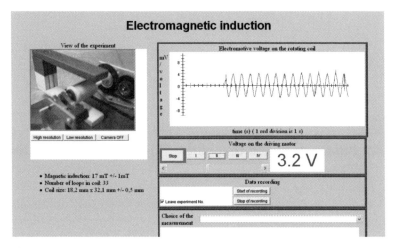

Figure 4.54 _ The web page on the client computer of the RE "Electromagnetic induction"; live web camera view, frequency controls and the graph with the time representation of the electromotive voltage (http://kdt-20.karlov.mff.cuni. cz/ovladani_2_en.html)

a homogeneous magnetic field, where the change of the magnetic flux is due to the coil orientation change with the time in an otherwise stationary magnetic field. The RE webpage of the experiment "Electromagnetic induction" is shown in Figure 4.54.

A) Lecture – As the introductory quantity for the electromagnetic induction field, we introduce the quantity magnetic flux Φ across the flat area described by the vector S in the homogenous magnetic field of induction B

$$\Phi = \mathbf{B} \cdot \mathbf{S} = BS \cos \alpha, \tag{4.40}$$

where α is the angle between vectors B and S (Figure 4.55).

In general, the magnetic flux may change due to:
- The time change of the value $B(t)$ (as in case of moving magnet inside pick-up coil),
- The time change of the area $S(t)$,
- The time change of mutual orientation of B and S, i.e. time dependence of the angle $a(t)$.

Electromotive voltage ε, $[\varepsilon]$ = V (volt), induced in the single turn due to Faradays' law of electromagnetic induction is

$$\varepsilon = - \frac{d\Phi}{dt}. \tag{4.41}$$

If we take the simple case of the steady state rotation by the constant angular velocity ω

$$\omega = 2\pi f = \frac{2\pi}{T}, \tag{4.42}$$

where f is the frequency and T the period of the motion, then for the electromotive voltage it is

$$\varepsilon = - BS \frac{d \cos (\omega t)}{dt} = BS\omega \sin (\omega t). \tag{4.43}$$

From the expression 4.43 we can see the increasing amplitude of the generated electromotive voltage with the increasing angular velocity ω of the rotation, obvious also from Figure 4.56.

Figure 4.55 _ Single turn rotating in homogeneous magnetic field

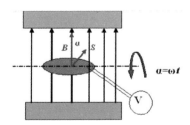

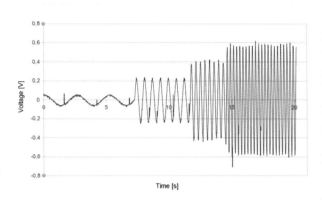

Figure 4.56 _ Electromotive voltage in rotating turn for increasing angular velocity (from left to right)

For a deeper insight into Faradays' law, we may discuss the circumstances of the electromotive voltage generation in a straight wire moving in the homogenous magnetic field with the students. The question may be:

"What is the value of the induced electromotive voltage in a straight wire of length L oriented in east-west direction, falling free in the Earth's magnetic field?"

B) Seminar – Students first prove electromagnetic induction phenomenon using simulation by W. Fendt "Generator" shown in Figure 4.57, (http://www.walter-fendt.de/ph14e/generator_e.htm).

For example, they may notice the dependence of the instantaneous coil position and the induced voltage or may change the direction of coil rotation.

Figure 4.57 _ Interactive simulation "Generator" http://www.walter-fendt.de/ph 14sk/generator_sk.htm

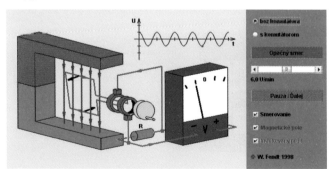

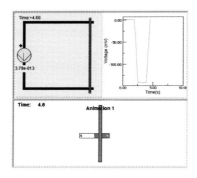 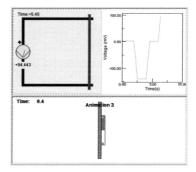

Figure 4.58 _ Example of simulation of electromagnetic induction due to permanent magnet moving in a coil http://www.compadre.org/physlets/electromagnetism/prob29_7.cfm

Seminar 7 – Electromagnetic field

Firstly, work with simulation to Problem 1. Fendt

A wire moving in the Problem 1. Problems 2 & 3 magnetic field, Il 29.2

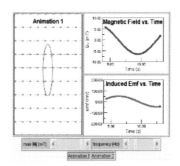

1. A flat circular coil with $N = 50$ turns, a radius of $r = 0.1$ m, is rotating uniformly with the frequency $f = 30$ Hz around its axis perpendicular to the magnetic induction $B = 10^{-3}$ T of the homogeneous magnetic field. Find the time dependences of the magnetic flux and electromotive voltage.

2. Horizontally oriented metal rod of length $L = 2$ m starts falling with zero initial speed in the gravitational field of the Earth. Find the voltage between the ends of the rod at time $t = 3$ s, when the initial orientation of the rod is N-S (W-E).

Figure 4.59 _ Example of the seminar assignment on the Electromagnetic induction in LMS MOODLE

As an application of the electromagnetic induction, the students may show the differences between the function of a dynamo and of an alternator and other interesting daily life applications.

An example of the seminar exercise is shown in Figure 4.58, where the task is to predict the sign of the voltage while moving the permanent magnet in four of its different possible orientations [81].

Electromagnetic induction assignment for the seminar exercise, taken from the MOODLE course in Electricity and Magnetism, is shown in Figure 4.59.

C) Laboratory exercise – Here we focus on the experimental verification of Faraday's law in two situations of the rotating coil in the homogeneous magnetic field (Figure 4.54) and free fall moving permanent magnet (Figure 4.53).

In the case of the rotating coil we obtain, by integrating the electromotive voltage ε from Faradays = law (equation 4.43), for one swing o-$T/2$

$$\int_0^{T/2} |\varepsilon| \mathrm{d}t = \int_0^{T/2} BS \sin(\omega t)\mathrm{d}t = 2BS = const., \qquad (4.44)$$

which is accessible to the experimental verification from the data obtained from the measurement. The extract from the students' laboratory project with the goal of proving this fact using the graphical program Origin, is shown in Figure 4.60. In the upper part of the figure is the time dependence

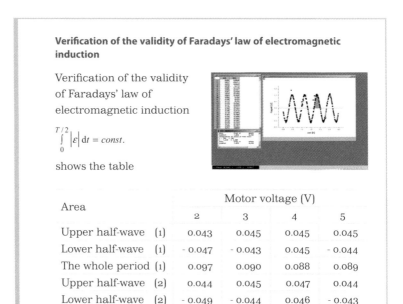

Verification of the validity of Faradays' law of electromagnetic induction

Verification of the validity of Faradays' law of electromagnetic induction

$$\int_0^{T/2} |\varepsilon| \, dt = const.$$

shows the table

Area		Motor voltage (V)			
		2	3	4	5
Upper half-wave	(1)	0.043	0.045	0.045	0.045
Lower half-wave	(1)	- 0.047	- 0.043	0.045	- 0.044
The whole period	(1)	0.097	0.090	0.088	0.089
Upper half-wave	(2)	0.044	0.045	0.047	0.044
Lower half-wave	(2)	- 0.049	- 0.044	0.046	- 0.043
The whole period	(2)	0.093	0.089	0.093	0.087

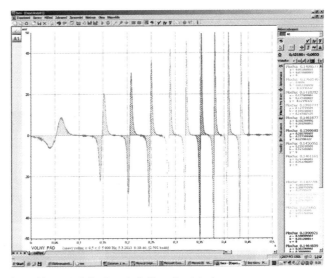

http://remotelab4.truni.sk/faraday.html.

Evaluation of integrals according equation 4.44 by ISES

Coil	B.S [T m²]	
	Upper swing	Lower swing
1	0.126	0.141
2	0.142	0.139
3	0.146	0.140
4	0.146	0.146
5	0.142	0.140
6	0.143	0.146
7	0.143	0.139
8	0.146	0.139
9	0.142	0.136
10	0.140	0.135
Average value	0.138	0.140

Figure 4.60 _ Extract from students' laboratory work – verification of Faradays' law by means of the rotating coil (up) andre "Free fall" (down) measurements and integrated by ISES

of the voltage taken from the rotating coil. The shaded area represents the integral (equation 4.44), and evaluations for different frequencies of rotation (given by the voltage on the driving motor), which are collected in

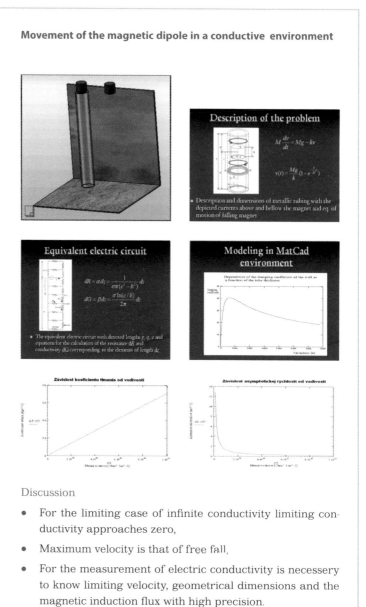

Figure 4.61 _ Sample of students project from subject matter of magnetic dipole motion in conductive media

the table. It is obvious that the constant value of the integral (4.44) is fulfilled.

The second example for the elucidation of Faradays' law is the falling magnet experiment. The record of the time dependence of the induced electromotive voltage in pick-up coils, is shown in Figure 4.60 together with the evaluation according to the equation 4.44.

The measurements were carried out by ISES system using the procedure for the calculation of definite integral for the time interval <0, $T/2$>. ISES system resolves individual integrations by colour. In the table are shown the results collected proving the applicability and verification.

D) Students' work competition – The study of magnet motion in conductive metal tubes was the topic of an advanced student project of the Department of Physics, Faculty of Education at Trnava University in Trnava "Motion of magnetic dipoles in electrically conductive media". The results of this work dealing with the modelling of the falling velocity of the magnet in a conductive tube in MatCad environment are presented in Figure 4.61.

4.8 Research-based teaching – "Photon theory"

The photon theory created at the beginning of 20th Century was one of the most radical revolutions in the field of physics. Behind this development were some experiments which brought enormous controversy in the process of their explanation by classical theories. These were Black body radiation and its explanation by Max Planck, Photoelectric effect explained by Albert Einstein and the Compton observation of electromagnetic waves scattering, topped by the doctor students, C. J. Davisson and L. Germer, when measuring the reflectivity of electron from metal surfaces. All these works testified to the character of waves and to the world of micro particles. It is of the utmost importance for all students of natural sciences and technology to get acquainted with those basic experiments. With respect to the price, the demand of their provision to the interested via remote experiments is obvious.

We exerted much activity in the co-operation of Charles University in the setting-up of the REs covering the quantum phenomena, among them "Photoelectric effect". We offer this RE in a laboratory exercise along with their hands-on and virtual counterparts. Let us describe RE "Photoelectric effect" and its experimental results.

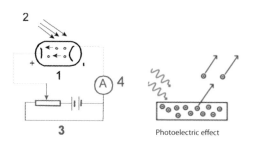

Photoelectric effect

Figure 4.62 _ Schematic arrangement of Photoelectric effect experiment – photon tube with photocathode and anode (1), electromagnetic radiation source (photons) (2), voltage source (3), A-meter (4)

The principle of the photoelectric effect rests in the emission of photoelectrons from the solid-state surface when illuminated by electromagnetic radiation. The great and observed contradiction was the way of obtaining energy from radiation not continuously, but in quanta of energy photons. If the energy of a photon is equal or exceeds the energy necessary for an electron to be released from the surface, the electron escapes. The effect is called photoelectric effect and the process itself, photoemission. A schematic arrangement of the experiment is shown in Figure 4.62.

Einstein explained the photoelectric effect by discontinuous emission and spreading of the electromagnetic radiation quanta, later called photons. He gave the photon the energy (together with M. Planck)

$$E_f = hf, \tag{4.45}$$

where h is the Planck constant $h = 6{,}6256.10^{-34}$ Js and f is the frequency. The energy conservation equation of the photoelectric effect, called Einstein's equation, is then

$$hf = W + \frac{1}{2}mv^2, \tag{4.46}$$

where the energy of the photon is divided between the energy barrier, called work function W, due to the bonding and surface forces, and the kinetic energy of the released photoelectron. The straightforward consequence of the equation 4.46 is the absence of photoelectrons if the radiation has a frequency lower than

$$f_0 = \frac{W}{h}. \tag{4.47}$$

In this way, the major problem of the interpretation of photoelectric effect by the classical theory was solved. For the current suppression we may use the electric field with the corresponding stopping voltage U_s

$$eU_s = \frac{1}{2}mv^2, \tag{4.48}$$

where e is the electron charge ($e < 0$, $U_s < 0$), then the Einstein's equation 4.46 is also

$$hf = W + eU_s. \tag{4.49}$$

In Figure 4.63 is the setup of the laboratory experiment "Photoelectric effect" from the components of fy Phywe replenished by the ISES modules for voltage and current measurements.

The setup is composed from several devices. For the data collection and their graphical representation and evaluation by ISES, the DC source of the variable voltage (from –5 V to +5 V) was built in the phototube circuit (1). Other components are Xe discharge tube (2) and its source (3). The signal output from the amplifier (4) is fed to the ISES voltmeter.

The measurement of ampere-volt characteristics $I(U)$ is then a simple task for data collection (Figure 4.64). We replenished the experiment with

Figure 4.63 _ RE "Photoelectric effect" with ISES data colection, phototube (1), Xe discharge tube (2) with source (3) and signal amplifier (4)

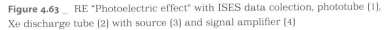
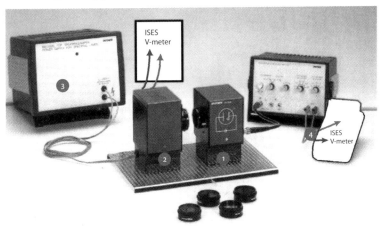

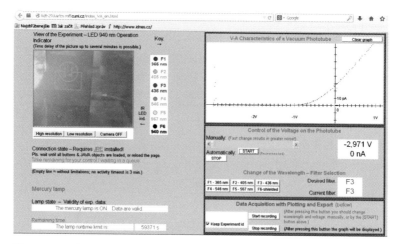

Figure 4.64 _ The web page of RE "Photoelectric effect" http://kdt-29.karlov.mff. cuni.cz/index_va_en.html

the manual filter holder (not depicted in Figure 4.63) with optical interference filters for different wavelengths.

Experimental basis for research activities in photoelectric effect may be RE http://kdt-29.karlov.mff.cuni.cz/index_VA_en.html or through http://www.ises.info from the co-operating Charles University in Prague. The web controlling page of the experiment is shown in Figure 4.64 [91].

An example of $I(U)$ characteristics measurements for three different wavelengths (580 nm red, 493 nm green, 365 nm blue) is shown in Figure 4.65.

Figure 4.65 _ Photoelectrics effect – $I(U)$ characteristcs

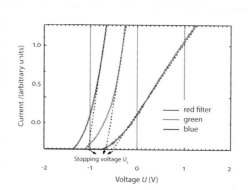

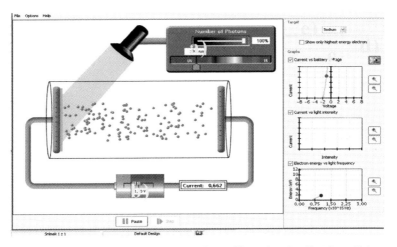

Figure 4.66 _ View of the interactive simulation Photoelectric effect (http://phet. colorado.edu/en/simulation/photoelectric)

From the characteristics it is obvious that with decreasing wavelengths the absolute value of the stopping voltage is increasing.

The equivalent to the above presented RE is the virtual experiment on photoelectric effect (http://phet.colorado.edu/en/simulation/photoelectric), the outstanding solution from the University of Colorado, in the frame of project Physics Education Technology (PhET), whose front page is shown in Figure 4.66. The "measurements" data from interactive simulation provide more or less identical data as a real experiment. The simulation is equipped with controls for several parameters, plotting the results into graphs and producing photo snaps of the graphs for further processing.

A) Lecture – RE "Photoelectric effect" is usually used for the introduction to quantum mechanics starting famous experiments. On opening the lecture, the teacher projects the experiment and shows its contradictory features in the light of classical theory of the 19th Century and discusses how they were resolved. The view of the lecture materials introductory quantum mechanics, taken from LMS MOODLE, is shown in Figure 4.67.

B) Seminar – Seminars in physics take place in a computer room, so the teacher may combine simulations and remote experiments across the Internet with classical exercises. Examples of the assignments with simulations and exercises are shown in Figure 4.68. We start with an interactive virtual experiment where students in a research-like process are strength-

Lecture 3

Photons-Energy and momentum

Compton effect- the photon collision with an electron

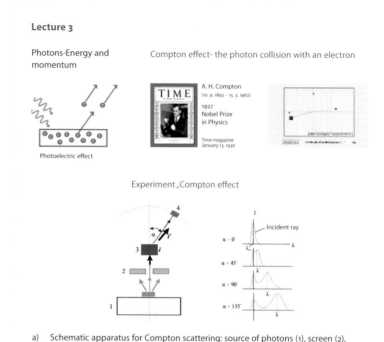

Photoelectric effect

A. H. Compton
(10. 9. 1892 – 15. 3. 1962)

1927
Nobel Prize
in Physics

Time magazine
January 13, 1936

Experiment „Compton effect

a)   Schematic apparatus for Compton scattering: source of photons (1), screen (2), scattered solid state matter occurs (3), photon detector (4)
b)   The scattered light intensity distribution at different scattered angles a

Figure 4.67 _ WWW page from LMS MOODLE course Physics II, Quantum mechanics introduction – photons, energy, mass and momentum, Compton effect

ened with the knowledge acquired from the lectures. The main focus is on the experiments already known from lectures and with the knowledge gained from those, are calculations and the solving of specific exercises started, as for example those shown in the assignment in Figure 4.68.

C) Laboratory exercise – Each student in the subject of Introductory Quantum Mechanics (PHYSICS II) takes laboratory exercises where he/she measures and delivers projects with the achieved results (Figure 4.69). Every student has to have his/her results signed by the teacher and also has to sign them as well (see the signed document in Figure 4.69 down, left).

D) Project – In research-based education of Quantum Physics, students often choose projects for achieving deeper insight, frequently as a continuation of the laboratory exercise. In Figure 4.70, the controlling webpage of

What is external fotoeffect? Try first and compare with calculations

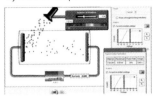

1. Medium wavelength of the lamps radiation of the metal fibre is $\lambda_m = 1200$ nm. Assuming 100 % efficiency, determine how many photons per second emits the bulb which electrical power is 200 W.

2. Cesium was irradiated with the light of wavelength $\lambda = 486$ nm and found the stopping voltage $U_b = -0.658$ V. Find out what the stopping voltage for the light wavelength $\lambda = 400$ nm. Compare with the simulation.

3. Photons of the wavelength $\lambda = 342$ nm are falling on the lithium surface and release photoelectrons that move on the circular orbit with radius $R = 1.2$ cm in the magnetic field of strength $H = 15$ Am⁻¹. Determine the work function W of lithium. See the simulation.

4. It was found in Compton scattering experiment that the scattered photon is deviated by an angle $a = 60\,°$ from the original direction while the electron describes a circle of the radius $R = 15$ mm in a magnetic field strength $H = 200$ Am⁻¹. Find the wavelength of the incident photon, if you know that the Compton wavelength of the electron is $\lambda_o = h/(mc)$ $= 2.426 \cdot 10^{-12}$ m.

Figure 4.68 _ Sample assignment for seminar on photons

the RE "Photoelectric effect" is shown, with the resulting curves after finishing the measurements of the project "Planck constant – the major physical constant of the Real World".

Closing the chapter devoted to our experience and practise of research-based education based on INTe-L strategy, presented in the form of eight topical case studies, we may conclude that the goals presented in paragraph 3.2 were met, predominantly by an increased initiative and the constructive approach of the students, reflected in improved results and the resulting marks achieved.

Assignment

- Acquaint yourself with the simulation of the University of Colorado PhET
    - Photoelectric effect http://zamestnanci.fai.utb. cz/~schauer/PhET-1.0/simulations/sims1dbo. html?sim=Photoelectric_Effect and
- Realize that the basic principles of the photoelectric effect are
    - An essential characteristic of the photon as an energy quantum of radiation,
    - The fact that the emission, starting from certain wavelengths, toward increasing wavelengths, does not occur,
    - The stopping voltage, and the work function W.
- From measurements of $I = I(U)$ characteristics of the phototube for selected wavelengths determine:
    - The value of the Planck constant h,
    - Work function W of the photocahode material,
    - Make a graph of $I = I(U)$ characteristic and find the stopping voltage U_s.
- Plot the dependence $1 / \lambda = f(U_s)$ and determine the value of Planck's constant using the modified Einstein equation

$$\frac{1}{\lambda} = \frac{W}{hc} + \frac{e}{hc} U_s.$$ and work function of the photocathode W.

Measurement

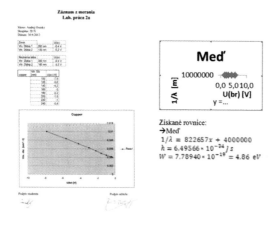

Figure 4.69 _ Sample of the students' laboratory assignment "Photoelectric effect" and selected measurements

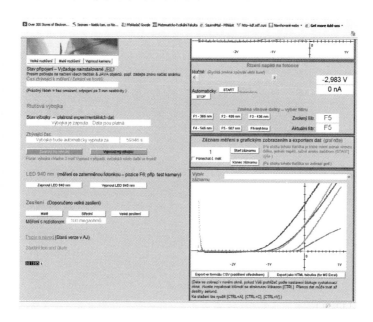

Voltage U [V]	Photocurent I [nA]	Wavelength λ [nm]	Resistor [MΩ]	Number of measurem. N	ΔU [V]	ΔI [nA]
0.032	0.0305	365	100	144	0.0013	0
-2.9724	0.0304	365	100	2	0.0037	0
-2.9553	0.0296	365	100	2	0.0110	0.003
-2.9638	0.0280	365	100	3	0.0024	0.003
-2.9589	0.0265	365	100	2	0.0098	0.003
-2.9574	0.0250	365	100	3	0.0131	0.003
-2.9566	0.0234	365	100	2	0.0074	0.003
-2.9541	0.0220	365	100	3	0.0123	0.003

Figure 4.70 _ View of the web page RE "Photoelectric effect" and measured data for the student's project "Planck constant – the major physical constant of the real word"

Chapter 5
Future of ISES remote laboratories

5.1 Integration of remote laboratories worldwide

The earlier era of remote laboratory development saw more efforts directed at their technical architectures [6][7], preoccupations including experimenting with technologies for real-time audio and video streaming in an effort to overcome bandwidth limitations while ensuring service quality, and dealing successfully with the handling of multiple simultaneous connections to shared online laboratory apparatus and equipment. A great deal of software (SW) and hardware (HW) approaches have been adopted (see the comparison of SW approaches to RE [49]. To a significant extent, many of these issues have been successfully overcome with continuous, reliable and high quality services being maintained for much of the past decade. The focus of RL development is now moving towards more sustainable models that promote promotes both institutional and individual engagement. Rather than individual academics custom building equipment for their specialized subjects, RL development and sharing is increasingly being carried out by multi-institution consortia like iLab [92], Library of Labs (LiLa) [93] or Labshare – Sahara [94].

In 2008, the Consortium of Czecho-Slovak Remote laboratories REM-LABNET was founded by Charles University in Prague, Tomas Bata University in Zlin and Trnava University in Trnava and announced the e-laboratory with at present 30 RE for universities and secondary schools [67] with examples from Trnava shown in Figure 5.1 and Prague laboratories in Figure 5.2. (The lists of existing RE of the Consortium REMLABNET is in Prologue Figure P1 and P2 and a complete description of individual experiments can be found in the Appendix).

The development is continuing, as there are numerous initiatives being either funded or proposed in order to create Remote Laboratory Management Systems (RLMS) that provide a common online portal for accessing and administrating a wide pool of heterogeneous developed remote lab systems that might be distributed at several universities. Examples include iLab (USA) [92] and Sahara (Australia) [94], with extension to UNED (Spain) [95]. This allows academics to take advantage of pre-existing tools to implement their experiments, rather than having to start from the very beginning. A next step would be exploiting such systems and implementing them at non-partner universities in order to get more feedback and consequently analyze the associated challenges with integrating different types of remote lab systems.

In the Czech and Slovak Republic, the situation in the creation of the nets of RLs has not progressed substantially. As documented in [67], about 30 RE have been built, mainly as the initiative of individual working places from different fields of technical and natural sciences disciplines. In Europe, compared to overseas initiatives, the activities in the creation of RL nets are only starting (see activities of the EU FP7 project Go-Lab [96]).

To contribute, the Consortium initiative REMLABNET arose to start research and realization of our own system of a Remote Laboratory Management System called REMLABNET for the segment of secondary schools and the lower level of university teaching. The project intends to help remedy the situation by the basic research initiative and suggests to carry out

Figure 5.1 _ WWW page of the remote experiment "Energy transfer in RLC circuits"; the frequency scan for the approximate determination of the resonant frequency [95] http://remotelab3.truni.sk

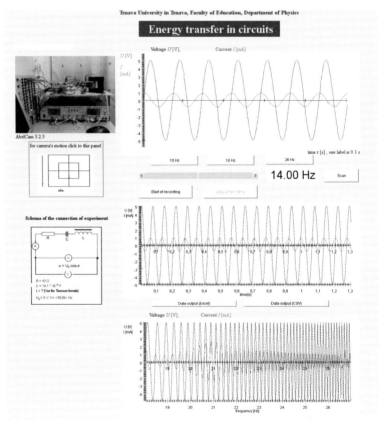

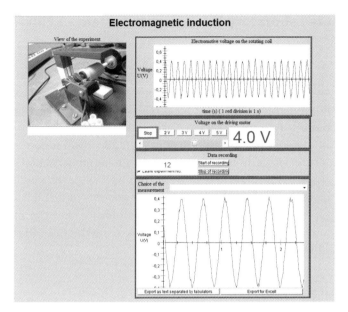

Figure 5.2 _ Web page of the remote experiment "Faraday's law of electromagnetic induction" http://kdt-20.karlov.mff.cuni.cz/ovladani_2_en.html

the research on the open system of the Remote Laboratory Management System (RLMS) and all its new constituent blocks, making use of all the advantages of virtualized Cloud computing [97]. The purpose of the project is thus to bring the research and testing together and introduce as a pilot project, the grid of inter-connected data centers under one RLMS for the management of remote laboratories for practical use of the participating Universities incorporating existing RL and RE. The system will be accessible for any interested e-laboratory free of charge, with access not being limited by the special interface used. Also, a wider participation of Czech and Slovak secondary schools is envisaged. The main goal is to enrich the teaching and learning process in natural sciences [98].

Next, we will describe the project REMLABNET in more detail.

5.2 Remote Laboratory Management System – REMLABNET

RLMS is an emerging trend which is receiving increased attention from many universities and being elaborated in the USA, Australia and Europe. The project highlights many new basic research aspects of integrating diverse remote lab systems associated with many challenges that are actually confronting each university which is willing to wrap a pool of existing remote laboratories in an RLMS. Extending the implementation of RLMSs among universities is associated with its own new research challenges owing to the diversity of the remote laboratory systems in terms of technical and pedagogical issues. We would like to emphasise that the main basic research outcomes of the envisaged project, are the final RLMS product as a whole and its four new vital components.

REMLABNET introduction

RLMS in general is a new system for the complex management of diverse interfaces of e-laboratories with manifold services, necessary for the proper function of a grid of e-laboratories. Due to this fact, a great deal of research and its results are supposed within the suggested project.

RLMS REMLABNET serves for the integration of REs from many e-laboratories covering many areas of natural sciences. In spite of the diverse interfaces used and various ICT applied, their inclusion in the system is supposed to be simple enough and ensure their availability to the general pool of interested clients. The system is running on modern facilities and data centers that enable Europe-wide integration with guaranteed 24/7 access. Multilingualism of all parts of the system is commonplace. The modular system enables the creation of virtual classes, which can exploit very sophisticated instruments, not available to the general public. The system will provide an easy way to store and access the resources for measurements, instruction and cooperative research across the EU. The main focus of the envisaged RLMS REMLABNET project is in the integration and management of simple experiments at university and secondary level and even at primary education levels, not covered worldwide.

The working principles of the RMLS are demonstrated in Figures 5.3 – Figure 5.5 – where its three main functionalities are clearly shown:

a) Server-clients connections through RLMS to one rig – Figure 5.3 describes the connection of three clients to one rig via RLMS. Control interfaces (web pages) are distributed via the web server in the RLMS to the clients and http protocol is used. The webpage is designed in general in HTML 5 using JavaScript widgets and/or PHP (Hypertext Preprocessor). Users connected to a single rig communicate mutually via the communication server, also hosted within RLMS. For this purpose, communication protocols like VOIP (Voice Over Internet Protocol) and RTSP (Real Time Streaming Protocol) will be used. Because the rig allows the control only to one client with the proper rights at a time, the other two clients can observe only the rig's activities. After the user with control rights has disconnected, the second connected user obtains them.

b) Server-clients connection to different rigs – Figure 5.4 depicts the second situation, i.e. the connection of multiple clients to different rigs, using RLMS. Control interfaces (web pages) are distributed via the web server of the RLMS to the clients. Because the users are not connected to identical rigs, mutual communication between users is not allowed. All clients have exclusive control of their respective experiments. Clients communicate via the standard TCP/IP (Transmission Control Protocol) socket communication. This solution of direct connection ensures fast transmission of control commands and measured data.

c) The server-clients connection through the RLMS to the experiment form-

Figure 5.3 _ Remlabnet function mode 1– multiple clients

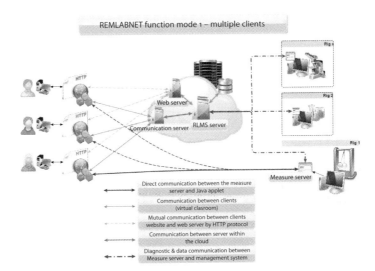

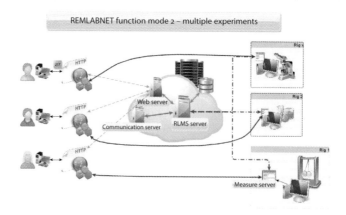

Figure 5.4 _ Remlabnet function mode 2 – multiple experiments

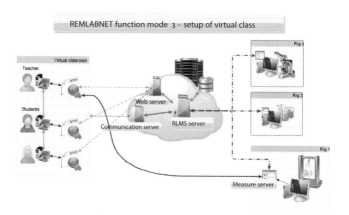

Figure 5.5 _ Remlabnet function mode 3 – setup of virtual class

ing the virtual class with a tutor is shown in Figure 5.5. Here it illustrates the connection of multiple clients to a single experiment by using a virtual classroom organized via RLMS. Control interfaces (web pages) are distributed via the web server in the RLMS to the participating clients. On the control webpage will be a special plug-in, which ensures the virtual classroom organization. Communication protocols like VOIP and RTSP will be used. The users connected to the virtual classroom communicate mutually via communication server, also resident in RLMS.

The individual rights (teacher/student) are assigned to the participating users. The teacher has the exclusive rights to control the experiment but he/she is allowed to pass control (or otherwise remove) to any of the participating students. The virtual classroom provides special features for the testing and evaluation of knowledge of the students.

Features of REMLABNET

Interface recognition and connection – one of the envisaged and highly required qualities of RLMS REMLABNET – is the necessity of recognizing a variety of common interfaces the e-laboratories may possess.

On applying to the RLMS, the system responds by the connection of the corresponding transformation driver, transforming a rig's data output to XML format. On the other hand, the system behaves in a bi-directional way, transforming the data sent by the client to the format of the rig. In such a way the system will by quite universal, removing all the communication barriers.

Registration of RE in RLMS – As a communication protocol for the data processing with the server JSON (JavaScript Object Notation) or XML (Extensible Markup Language) will be used. Once the connection is established, the system will send a crafted webpage to a web server and ensure experiment inclusion in the database with all the necessary information that the user enters during the experiment setup. This information should allow the inclusion of the experiment in the appropriate category, describing its physical background and its functioning. Descriptions will not be restricted to mere textual information, but will allow the inserting of images and videos to enable the achievement of dealing with the deepest possible problems with clear understanding. This step of the experiment is included in the database and available on the web portal REMLABNET.

- Virtual Classroom – This service will allow the integration of the rig and entry of the users into a virtual classroom. The virtual class setup will also enable (student/teacher) roles allocation. The virtual classroom will also provide special features for the testing and evaluation of the student's knowledge. Communication within the virtual classroom will be by the video conferencing or text-only chat. For this purpose communication protocols like VOIP and RTSP will be used.

- Communication board – This service will be a simple communication window that will serve to allow questions to be directed to the administrator or the insertion of proposals for improvements and feedback by email. This feature will be fully automated. Communication with the administrator will be displayed on the whiteboard window.

- Booking System – This service allows the booking of experiments a selected time and thus ensures exclusive access to the measurement. Reservations will be based on registration.

- Simulation inclusion – The remote experiments in the system can be rather complex and may contain many input parameters for their settings. The measurement cycle of such type of experiments can also be very long and therefore the setting of incorrect parameters will lead to a considerable loss of time. Therefore, our goal is to insert the corresponding simulation of the phenomenon in question on the basis of its mathematical formulation in to the measuring process. This service allows the user to simulate the process in question in advance and/or simultaneously with the measuring process with the variable input parameters. Thus, the user will achieve information by sensitive analysis about the influence of individual parameters that may help in the setting and data evaluation of parameters. Data obtained from the measurements and simulation will be depicted in graphs for comparison.

- Entrance test – This function will restrict access to the experiment only to those who have passed the test entered by the administrator. It will test the user's knowledge and prevent misuses of the experiment. The results of these tests will be stored and may be used for statistical purposes.

- Diagnostics and feedback – Diagnostic services should ensure keeping track of the status of the experiments connected to the RLMS. Depending on either the activity of the rig, its breakdown or failure, the status will be continuously displayed at the access portal of RLMS. Diagnostics should also allow sending commands to the experiment in case of detected faults. Because the RLMS will manage many experiments at one time, it is necessary that the interface will allow connection to and supervision of many rigs, and simultaneously to be able to identify the individual rigs and communicate with them individually. As there will be an IP address for each rig stored in the server database, it will be appropriate to use these addresses for identification. There will be measures which serve for the auto-repair of experiments. If the step does not bring the necessary remedy, the system will contact the rig's administrator automatically informing him/her about any errors found and requiring repair. The system will support not only the diagnostics of the experiment as a whole, but its parts as well, to relieve the supervising load of the rig's administrator as much as possible. From the readings transmitted from the individual sensor, the system will determine its malfunction, giving the administrator the clue for its repair.

• Transformation of rig's data to XML – The main purpose of this transforming interface is to gather experiment configuration, current controlling values and measured data results in order to maintain a particular experiment setup and reproduce it off-line. Every rig is planned to be equipped with a separated storage space which should include the date, time, logged user and description, identifying the experiment and the measured data gathered from the experiment physical hardware (apparatus). This interface, giving data in standardized XML, will be used for a simulation process.

• Database storage – The system will provide storage of all information concerning experiments and enable their displaying in the form of a catalogue. In the database there will be measured experimental data of registered users stored for later use.

Parts of REMLABNET

MeasureServer (MS) – This is a vital component of every RE, ensuring its proper time sequence functioning. It is a state machine [53], provided with a comparatively small psc file. MS manages all communication with the hardware and works on the basis of instructions, contained in a psc file. It is a very powerful tool enabling the control of any rig. It avoids the necessary programming of a special MS for every physical HW.

ER ISES – For each experiment the logic of its operation has to be programmed. This shows the sequence of conditions and actions of how that experiment is carried out. For example, at which moment during the experiment does the measuring procedure start, or vice versa, at which moment during the experiment does the measuring procedure stop? The creation of this control logic requires programming at an advanced level. Our goal here, as part of the basic research, is to develop a graphical development environment that automates this programming process and compiles the control code for users who are not expert programmers. For the set-up of a new RE it is also necessary to ensure the generation of control interface (website) for the client. This activity is automated as well, by developing the graphical development environment for the webpage generation [99].

Web server and web space management – The aim is to create a web space based on the HTML5 CANVAS as the e-table. This will be implemented with JavaScript, Box2D, representing the laws of physics in real-time on the screen.

RLMS includes a portal (virtual space) created by the web tools with the components e-work, e-table, e-research, e-databases and e-library, designed using the latest and most advanced methods and means like HTML5, JavaScript, AJAX or others. The HTML5 environment will ensure portability and compatibility for a wide range of devices such as mobile phones, tablets, computers and many more. The environment will meet the need for easy work in the lab and work scope with attempts to meet new trends using the drag & drop technique. The whole environment will be represented in the form of dynamic webpages and by means of which all functions of the RLMS Web 2.0 environment will be accessible.

Data Warehouse – The data warehouse (DW) is the part of the system for storage and data analysis. It is a centralized data store, providing analysis services to MS, web server, image server and other components of RLMS. DW is a central unified system of SW database services in remote laboratories used for storage and data analysis. This system includes a number of sophisticated instruments providing data analysis from individual rigs. The data analysis may detect and filter noise or measurements errors.

Booking and management server – The reservation server generates a service enabling individual RE reservation for a given time period. The service will include not only booking services, but also services for the gateway interface in the form of SMS messages, containing information as a calendar, tasks meetings and to work in teams. The Service Management server is continuously checking the proper functioning of the MeasureServer and individual experiments.

Communication server – This is a system designed for the transmission of information and real-time communication, interaction and collaboration in the teaching and learning process with RE. The communication server will provide insertion of RLs in social networks and allow the working of experiments within groups and enable mutual help among the groups. The communication server will also serve as the interface for connecting to both the internal knowledge base, as well as to the global knowledge base such as Wikipedia. Part of the communication server will be the authentication interface with a direct connection to the data of stored projects.

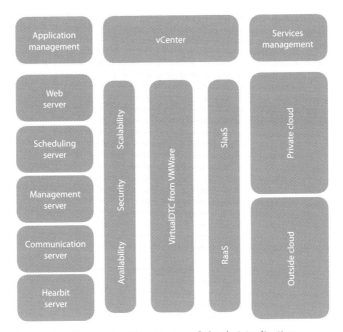

Figure 5.6 _ Schematical functioning of cloud virtualisation

Virtualized Cloud – (Schematic functioning of virtualization is shown in Figure 5.6.) To provide optimal access to all the experiments and economical exploitation of the RLMS with all its functionalities and benefits, we intend to use virtualized cloud computing. Our idea is to make two or more datacenters (DTC) with relevant data and clients connected to the nearest DTC with the lowest traffic and utilization. Because a large number of inexperienced users may access RLMS, we need to create a network environment which will be as secure as possible. Our specific concern is the RE computer system, the software from the instrument vendor, and the security of the data collected [100].

For this task, a few servers have been dedicated in two data centers. The vendors of these servers are Oracle (SunFire) and Cisco (UCS). All servers use the ESXi operating system from VMWare and one centre is dedicated to the management of all of them. The operating systems for each VMs are used specifically on purpose (for example MS Windows desktop or server editions, Linux SLES for VMWare or Ubuntu etc.).

The security of the DTCs is built on several levels. The first security level is based on IP (Internet Protocol) with ports and IP addresses which are either authorized or not authorized. The second security concern will deal with the unauthorized access to the instrumentation and computer desktop. Each user level is accessed using encrypted usernames and passwords und uses SSL to transmit between the DTCs. The third level is IDS (intrusion detection system) and IPS (intrusion prevention system) for monitoring possible incidents. The fourth and final level is SIEM (security information and event management) and the Checkpoint application firewall for monitoring the application level.

Remote laboratory in Trnava University in Trnava

Introduction

After two pilot years, the first Slovak natural sciences e-laboratory of remote experiments was opened with an opening ceremony on 30. 09. 2010 in Trnava University and is freely accessible 24/7 from the URL http:// kf.truni.sk/remotelab. It is located in the premises of the Department of Physics, Faculty of Education, Trnava University in Trnava, Slovakia. All the remote experiments are built using the physical hardware ISES and the environment for controlling programs and webpage compiling ER-IES (more details in Chapter 2). The experiments are freely accessible across the Internet using an arbitrary web browser. The communication language for the experiments is JavaScript, enabling the accessibility of RE via any communicators from smarthphones or book readers and ipads independently of their operating system.

At present, the e-laboratory in Trnava provides a set of remote experiments from the physics and chemistry and environment education. The experiments are freely accessible at http://remoelabN.truni.sk, where N = 1, 2, ..., 13. The e-laboratory is constantly enriched by new RE, e.g. in 2014 by the experiments "Joule experiment" and "p-V diagram of gases".

Figure A1.0 _ View of laboratory of remote experiments in Trnava University

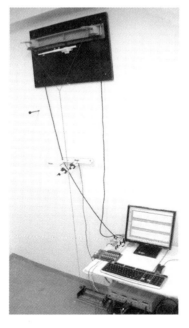

Brief description and use of individual remote experiments in Trnava e-laboratory

Remote experiments in environment monitoring

1.1 RE "Environmental monitoring in Trnava" (see Figure A1.1)

Lab type: Remote experiment

Lab owner: F. Schauer, M. Ožvoldová and L. Tkáč

Grade level: Primary education (7-15 years old), Secondary education (15-18 years old)

Language: Slovak, Czech, English

Difficulty level: Basic

Interaction level: Data logger with data storage and data mining

Booking required: No

Keywords: Temperature, barometric pressure, sun irradiation, radiation background, data outputs

Web link: http://remotelab1.truni.sk,

Figure A1.1 _ WWW frontpage of the remote experiment "Environmental monitoring in Trnava" http://remotelab1.truni.sk

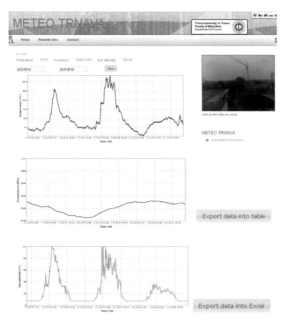

Brief description of the remote experiment: This experiment is the data logger of important environmental data – the temperature, the barometric pressure, the sun irradiation, and in the near future, the γ radiation background. The data is arranged by years, months, days and hours.

Description of use: First, the data may be used with an advantage for environmental observations locally or in comparative studies with geographically distributed similar weather stations. The second use is for acquisition of skills in work with data, data arranging in tables, plotting in graphs, data filtering and processing, finding average and extreme values in graphs, correlations etc. The remote experiment may be exploited as an advantage with seminars, project work, homework and examinations.

Remote experiments in Mechanics

1.2 RE "Simple pendulum" (see Figure A1.2)

Lab type: Remote experiment
Lab owner: F. Schauer
Grade level: Lower secondary education (12-15 years old), Higher secondary education (15-18 years old), University starting level

Figure A1.2 _ WWW front page of the remote experiment "Simple pendulum" http://remotelab5.truni.sk

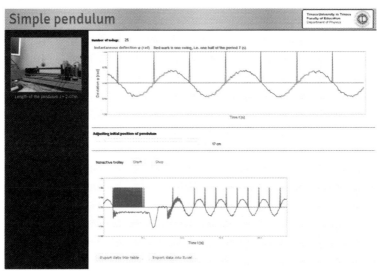

Language: English and Slovak
Difficulty level: Easy
Interaction level: High
Booking required: No
Keywords: Simple pendulum, kinematics, dynamics and energy in curvilinear motion, acceleration due to the gravity
Web link: http://remotelab5.truni.sk
Related simulation: http://www.walter-fendt.de/ph11e/pendulum.htm
http://www.aldebaran.cz/applets/fy_kyvadla/start.html

Brief description of the remote experiment: This is a classical simple pendulum with an adjustable amplitude and low damping, enabling experiments with duration of approximately 1 hour. The output is in the form of the instantaneous angle of deflection (red curve in Figure A 1.2) on time and the counter of the number of swings (blue marks). All data is available for downloading.

Description of use: The remote experiment is a simple pendulum with multipurpose use, starting from the measurement of acceleration of gravity, kinematics (instantaneous velocity and acceleration with its tangential and normal components), dynamics (acting forces, equation of motion), and energy (potential and kinetic energy, energy conservation law, work of dissipative forces) of curvilinear motion. The experiment is suitable as a demonstration experiment in the lectures, in the seminars as the data source for calculations and in the laboratory exercises.

1.3 RE "Free fall" (see Figure A1.3)

Lab type: Remote experiment
Lab owner: F. Schauer, M. Ožvoldová
Grade level: Lower secondary education (12-15 years old), Higher secondary education (15-18 years old), Bachelor course of physics
Language: English and Slovak
Difficulty level: Easy, basic
Interaction level: High
Booking required: No
Keywords: Motion in the gravitational field, free fall in vacuum, acceleration due to the gravity, conservative field
Web link: http://remotelab4.truni.sk/position.html

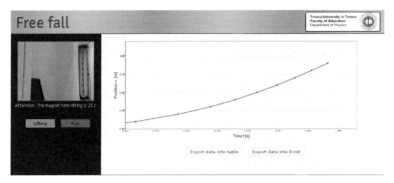

Figure A1.3 _ www front page of the remote experiment "Free fall";
web: http://remotelab4.truni.sk/position.html

Related simulation: http://www2.swgc.mun.ca/physics/physlets/mars_fall.
html,

Brief description of the remote experiment: The motion of a body (permanent magnet) in the gravitational field either damped by the resistance of the air or in the evacuated tube. The motion is detected by the equidistant coils and the instantaneous position is displayed. The starting position of the magnet is accomplished by the magnetic vessel lift. The remote experiment may be run in two regimes – one in the presence of the air and the second with the evacuated tube. (Also the signal generated due to the electromagnetic induction may be available for the study of electromagnetism – see further).

Description of use: The motion by the force of gravity is the omnipresent phenomenon for any physics course. It may serve for the demonstration in kinematics, dynamics and energetic of accelerated motions with constant acceleration, its dynamics and the mechanical energy conservation in conservative fields.

1.4 RE "Archimedes' principle" (see Figure A1.4)

Lab type: Remote experiment
Lab owner: M. Ožvoldová
Grade level: Lower secondary education (12-15 years old), Higher secondary education (15-18 years old), Bachelor basic physics course

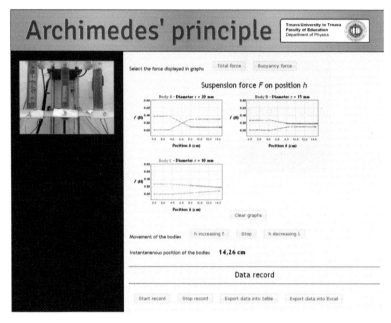

Figure A1.4 _ WWW front page of the remote experiment "Archimedes' principle"; web: http://remotelab9.truni.sk

Language: English and Slovak
Difficulty level: Medium
Interaction level: High
Booking required: No
Keywords: Buoyancy force, relation to liquid density, floating,
Web link: http://remotelab9.truni.sk
Related simulation: http://phet.colorado.edu/en/simulation/buoyancy

Brief description of the remote experiment: This experiment is based on measuring the suspension force when a body is immersed in a liquid. It uses three independent vessels with three dynamometers and moving platforms. From the results of the experiment, the weight of the body, the weight reduced by the buoyancy force and the net buoyancy force as a function of the immersion are accessible.

Description of use: The Archimedes' principle is a generally well-known physics law, but often without proper physical knowledge behind the buoyancy force. This experiment is carried out with an advantage in lectures as a demonstration of phenomena like atmospheric pressure or the floating of bodies in a liquid and in seminars for the calculations and in the laboratory exercise for the determination of the physical and geometrical parameters of bodies and liquids.

1.5 RE "Joule experiment" (see Figure A1.5)

Lab type: Remote experiment under construction (to be completed in 2015)
Lab owner: F. Schauer
Grade level: Lower secondary education (12-15 years old), Higher secondary education (15-18 years old), Bachelor courses of physics
Language: English and Slovak
Difficulty level: Medium
Interaction level: High
Booking required: No
Keywords: Heat as a form of energy, transformation of mechanical energy into heat, work of dissipative forces, friction forces, heat in solid bodies, specific heat
Web link: not yet complete (in 2015 it will be at http://remotelab.17truni.sk)

Brief description of the remote experiment: The Joule's experiment played a decisive role in presenting heat as a form of energy and the possibility of

Figure 1.5 _ View of the setup of the experiment "Joule experiment"; under construction http://remotelab17.truni.sk

1 – Hands on experiment, 2 – ISES interface, 3 – PC (ISES)

1 – Dynamometer, 2 – Voltage source, 3 – Roller,
4 – Optical gate, 5 – Thermometer, 6 – Driving motor (12 V; 2.2 A),
7 – Connecting wires, 8 – Weight, 9 – Plastic tape

transforming mechanical energy into heat. It was one of the first insights into the micro-world of physics' phenomena. The experiment relies on the transformation of the known work of friction forces into heat, observed by the temperature increase of various bodies with known thermal properties.

Description of use: This is one of the basic experiments for making a bridge between the phenomena from the macro to the micro world. It may be used either for proving the equality of macro and micro energy concepts. In reverse, it may be used for measuring the specific heat of solid bodies. The redundancy of the two independent units for energy and heat is easily shown by the experiment. It is useful for demonstration, seminars and laboratory exercises, but also for project and homework (when in remote form).

1.6 RE "*p-V* diagram of gases" (see Figure A1. 6)

Lab type: Remote experiment under construction (to be completed in 2015)
Lab owner: F. Schauer
Grade level: Lower secondary education (12-15 years old), Higher secondary education (15-18 years old), Bachelor courses of physics
Language: English and Slovak
Difficulty level: Medium
Interaction level: High
Booking required: No

Figure A1.6 _ View of the setup of the remote experiment "p-V diagram of gases" under construction (2014) http://remotelab19.truni.sk

Keywords: changes of state of the gas, adiabatic change, isothermal change, isochoric change

Web link: not yet complete (in 2015 it will be at http://remotelab19.truni.sk)

Brief description of the remote experiment: The equation of state for gases is the basis of thermodynamics and stems from the conservation of energy law. The remote experiment serves for measuring the changes of state of the gases and the present setup of the apparatus is used for the study of adiabatic, isothermal and isochoric effects. The experiment includes the output of data.

Description of use: The experiment is useful as both the laboratory experiment and for project work as well as homework. Experiments, integrating several typical changes of the states of gas, are rare, so the RE is uniquely suitable for verifying the behaviour of the gas, in lectures and seminars. In the laboratory exercise, it may be used for the determination of the Poisson constant of the adiabatic change of state, measuring the temperature in isothermal change and proving the temperature dependence on pressure in the isochoric change of state.

Remote experiment in electricity, magnetism and electromagnetism

1.7 RE "Energy transfer in RLC resonance circuits" (see Figure A1.7)

Lab type: Remote experiment

Lab owner: L. Tkáč and F. Schauer

Grade level: Lower secondary education (12-15 years old), Higher secondary education (15-18 years old), Bachelor courses of physics

Language: English and Slovak

Difficulty level: Medium

Interaction level: High

Booking required: No

Keywords: Transfer and phase characteristics of resonance RLC circuits, phase shift on individual RLC components, transfer of energy from the source to the resistive load

Web link: http://remotelab3.truni.sk/transfer_energy.html

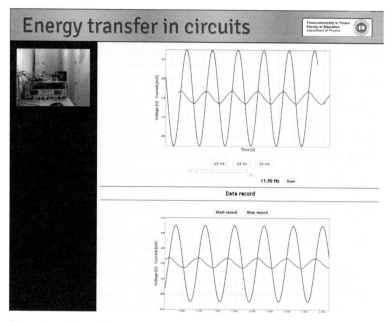

Figure A1.7 _ View of the web page of the remote experiment "Energy transfer in RLC circuits" http://remotelab3.truni.sk/transfer_energy.html

Brief description of the remote experiment: Frequency dependent phenomena in RLC circuits are the basis for both demonstration of phenomena in electricity and electromagnetism and for circuits with resonance phenomena. Usually, the voltage transfer and phase characteristics (i.e. dependence of both amplitude and phase on the driving frequency) are examined as the function of frequency, the assignment here is the transfer of energy from the source to the load (as in communication devices).

Description of use: Here, the current, voltage and corresponding energy relations of both electric and electromagnetic phenomena in passive circuit components – capacitor and inductor and corresponding RLC circuit properties, may be studied. This remote experiment may be used both as a laboratory experiment and also for project and homework.

1.8 RE "Phase in RLC circuits" (see Figure A1.8)

Lab type: Remote labs
Lab owner: L. Tkáč and F. Schauer
Grade level: Lower secondary education (12-15 years old), Higher secondary education (15-18 years old), Bachelor courses
Language: English and Slovak
Difficulty level: Medium
Interaction level: High
Booking required: No
Keywords: Phase in RLC resonant circuits, phase and amplitude characteristics of resistor, capacitor and inductor
Web link: http://remotelab3.truni.sk/phase_rlc.html

Brief description of the remote experiment: This is a modification of the previous experiment, where the phase relations for individual components, i.e. resistor R, inductor L and capacitor C may be studied with respect to the frequency.

Figure A1.8 _ View of the web page of remote experiment "Phase in RLC circuits" http://remotelab3.truni.sk/phase_rlc.html [instantaneous voltage (blue) and current (red) for the inductor depicted]

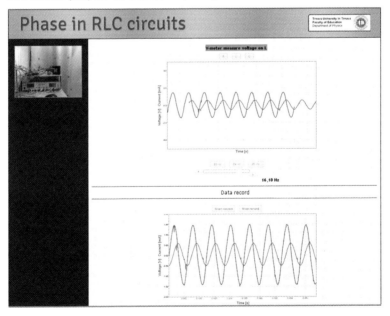

Description of use: This remote experiment may be used both as a laboratory experiment and for project work and homework. The individual components of the circuit may be determined via changing the damping. It may be used for the demonstration of the behaviour of passive RLC circuits in all forms of teaching.

1.9 RE "Transient phenomena in electric oscillators"
(see Figure A1.9)

Lab type: Remote labs
Lab owner: L. Tkáč and F. Schauer
Grade level: Higher secondary education (15-18 years old), Bachelor courses
Language: English and Slovak
Difficulty level: Medium
Interaction level: High
Booking required: Yes
Keywords: RLC circuits, response to perturbation in time domain, damping
Web link: http://remotelab6.truni.sk

Figure A1.9 _ View of the setup of the experiment "Faradays' law of electromagnetic induction " by falling magnet in the glass tube with pick up coils, the time dependence of the induced voltage in coils is depicted
http://remotelab4.truni.sk/faraday.html

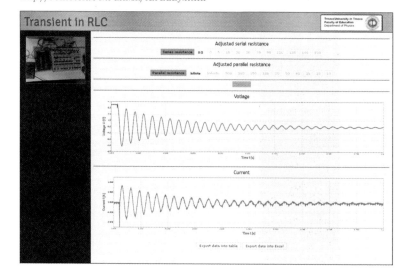

Brief description of the remote experiment: The electric and electromagnetic phenomena in RLC circuits with variable damping may be studied in the time domain as the response to the voltage perturbation. By changing the damping both by series resistance to the inductor and parallel resistance to the capacitor, we may find all the values of the components of the RLC circuit.

Description of use: This experiment is suitable for the determination of the natural frequency of the circuit, damping and RLC components of the circuit. The remote experiment may be used as a laboratory exercise, project and laboratory work.

1.10 RE "Characterization of the DC source of electromotive voltage"(see Figure A1.10)

Lab type: Remote labs
Lab owner: L. Tkáč and F. Schauer
Grade level: Lower secondary education (12-15 years old), Higher secondary education (15-18 years old), Bachelor courses
Language: English and Slovak

Figure A1.10 _ View of the web page of the remote experiment "Characterisation of DC source of electromotive voltage" http://remotelab7.truni.sk (loading characteristics – voltage (blue) and current (red) both on loading resistance)

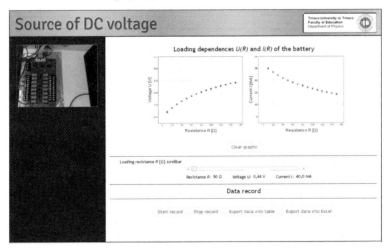

Difficulty level: Medium
Interaction level: High
Booking required: Yes
Keywords: DC sources of electromotive force, parameters of DC sources, characterization of DC sources.
Web link: http://remotelab7.truni.sk (see Figure A1.10)

Brief description of the remote experiment: The electrochemical sources of electric energy are important components for various appliances. Two basic parameters of any DC electromotive voltage source are important – the electromotive voltage and the internal resistance of the source. To determine those parameters, the method of loading characteristics is used, loading the source with variable loading resistors of known value. Both parameters of the DC source may be determined from the running of both dependences of voltage and current.

Description of use: DC sources of energy are omnipotent and very useful devices, whose parameters are often needed for practical exploitation. The RE is useful for both elucidation of the behaviour of the DC source and determining their crucial parameters. The RE is useful for both demonstration purposes and for homework and project work.

1.11 RE "LEDs radiation emission" (see Figure A1.11)

Lab type: Remote labs
Lab owner: L. Tkáč and F. Schauer
Grade level: Lower secondary education (12-15 years old), Higher secondary education (15-18 years old), Bachelor course of basic physics
Language: English and Slovak
Difficulty level: Medium
Interaction level: High
Booking required: Yes
Keywords: LEDs, material aspects of LEDs, wavelength of the LEDs emission, Plancks' constant
Web link: http://remotelab8.truni.sk

Brief description of the remote experiment: Electro-luminescence diodes are important semiconductor devices for transforming electric energy into radiation. The wavelength of the emitted radiation is inversely proportional

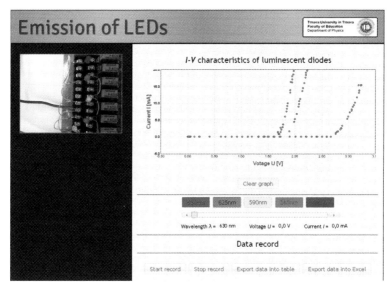

Figure A1.11 _ View of the web page of remote experiment "Leds radiation emission" http://remotelab8.truni.sk

to the energy band gap of the semiconducting material used, and that in turn is reflected in the turn-on voltage of the $I(U)$ characteristics. Thus, knowing the dependence of the wavelength of the emitted radiation on the turn on voltage of the LEDs, we may determine the Plancks' constant.

Description of use: This experiment may be used in general for the determination of the basic properties of electroluminescence diodes and for their wavelengths in particular. The remote experiment is useful as a laboratory experiment and project work and homework.

1.12 RE "Faraday's law of electromagnetic induction"
(see Figure A1.12)

Lab type: Remote labs
Lab owner: L. Tkáč and F. Schauer
Grade level: Lower secondary education (12-15 years old), Higher secondary education (15-18 years old), Bachelor courses of physics
Language: English and Slovak
Difficulty level: Medium

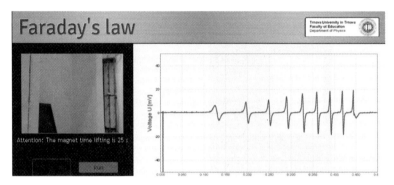

Figure A1.12 _ View of the web page of remote experiment "Leds radiation emission" http://remotelab8.truni.sk

Interaction level: High
Booking required: Yes
Keywords: Induced voltage, time dependent magnetic flux, electromagnetic induction, Faraday's law
Web link: http://remotelab4.truni.sk/faraday.html

Brief description of the remote experiment: Faraday's law is the basic law of electromagnetism with rich applications in daily life. It reflects the change of the magnetic flux with time, usually observable as electromotive voltage in a closed loop, here created by the moving permanent magnet, moving by a free fall, in the series of connected pick up coils. The signal depicts the instantaneous velocity of the fall, the growing velocity with time may be reconstructed and also the validity of Faraday's law, i.e. the constant charge passing in the circuit independent of the velocity may be studied.

Description of use: This remote experiment is useful for both the demonstration and verification of Faraday's law. Thus, it is useful for both demonstration purposes in the lecture and measurement with evaluation both in the seminars and laboratory exercise. The enormous advantage is the sampling of the signals in the 10 µs range, as the whole experiment takes fractions of seconds only.

Remote experiments in chemistry education

1.13 RE "Electrochemical cell" (see Figure A1.13)

Lab type: Remote labs
Lab owner: Z. Gerhátová and F. Schauer
Grade level: Lower secondary education (12-15 years old), Higher secondary education (15-18 years old), Bachelor courses
Language: English and Slovak
Difficulty level: Medium
Interaction level: High
Booking required: Yes
Keywords: Electrochemical sources of voltage, redox processes, Nernst equation
Web link: http://remotelab2.truni.sk

Brief description of the remote experiment: Electrochemical batteries, as sources of electric energy, are important components for various appli-

Figure A1.13 _ View of the setup of the experiment "Electrochemical cell" (left) and measured dependences of the electromotive voltage (blue) and electrolyte conductivity on the ions concentration (red) http://remotelab2.truni.sk

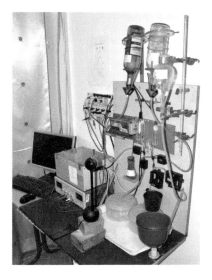
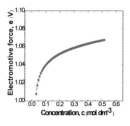
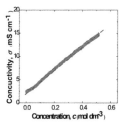

ances we use in our daily life. However, the functioning of such a basic device is not very common. This was the reason for building this remote experiment, where the variable is the concentration of ions, taking part in redox reactions, resulting in the electrochemical electromotive force and voltage. Both the electromotive force due to Nernst equation and the electrolyte conductivity on the ions concentration is measured during titration.

Description of use: This experiment is decisive of all the reactions and parameters of the electrochemical cell and their dependence on the electrolyte concentration. It may therefore be used for project work and homework.

New experiments under construction

1.14 RE "Incline " (see Figure A1.14)

Lab type: Remote labs
Lab owner: L. Tkáč
Grade level: Higher secondary Education (15-18 years old), University level
Language: English and Slovak
Difficulty level: Medium
Interaction level: High
Booking required: Yes
Keywords: Incline plane, energy concept of motion with dissipation, remote experiment
Web link: http://remotelab11.truni.sk

Figure A1.14 _ View of the web page of the experiment "Incline" (left) and setup of the experiment (right) – http://remotelab11.truni.sk

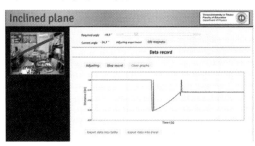

Brief description of the remote experiment: This is a multipurpose experimental kit for the incline building – INCLINE – for both the hands-on and the remote experiment versions. In connection with the ISES measuring system and its modules (ISES sonar and ISES V-meter), it enables the setup of the experiment with a variable and measured incline and with continuous measurement of the position of the moving body (wooden block) with dissipative forces present. Firstly, the experiment may be used for schooling (of teachers e.g. in the accredited courses) in the building of both computer-oriented hands on and later for even remote experiments without any programming, using standard ISES. The ER ISES environment may be then used for producing a web page and remote experiment controlling programs without any programming for those who are not specialists in informatics.

Description of use: This experiment has numerous applications in the teaching of mechanics in kinematical, dynamical and energy approaches in solving problems of the real world with dissipative forces present.

1.15 Teaching room of RE "Incline " (see Figure A1.15)

Lab type: Remote labs
Lab owner: L. Tkáč, F. Schauer
Grade level: Higher secondary education (15-18 years old), University level
Language: English and Slovak
Difficulty level: Medium
Interaction level: High
Booking required: Yes
Keywords: Teaching room for remote teaching of the whole class, in and pre- service teacher instruction courses on RE construction and building using INCLINE kit and ER ISES.
Web link: http://remotelab12.truni.sk (N varied from 12 to 16).

Brief description of the remote experiment: The experiments are identical to 1.16 RE, only at present they form the equipment of the teaching room currently equipped with 6 (10 in the future) identical REs.

Description of use: The teaching room has manifold applications and is intended to be used for teaching the whole class at any school using an Internet connection and web browser only. The enormous advantage is the accessibility of the experiments without any cost and maintenance care,

Figure A1.15 _ View of the web page of the experiment "Incline" (left) and setup of the experiment (right) – http://remotelab11.truni.sk

space and running costs. The teacher may then focus on the experiment only and control as facilitator, the process of research-based teaching. The second envisaged application is to use the teaching room for in and pre-service teachers instruction courses on RE construction and building using the INCLINE kit and ER ISES.

References

[1] M. F. Aburdene, E. J. Mastascusa, R. Massengale, and L. P. Grayson "A Proposal for a Remotely Shared Control-systems Laboratory", *Frontiers in Education: Engineering Education in a New World Order*, 1991, pp. 589-592.

[2] B. M. Leiner, V. G. Cerf, D. D. Clark, R. E. Kahn, L. Kleinrock, D. C. Lynch, J. Postel, L.G. Roberts, and S.Wolff, *"A Brief History of the Internet"*, accessed : 21.12.2014 http://www.isoc.org./internet/history/brief.shtml.

[3] B. Aktan, C. Bohus, L. Crowl, and M. H. Shor, „Distance Learning Applied to Control Engineering Laboratories", *IEEE Transactions on Education*, Vol. 39, No. 3, August 1996, pp. 320-326.

and J. Henry, "Controls Laboratory Teaching via the World Wide Web", *ASEE Annual Meeting*, Washington, D.C., June, 1996. http://chem.engr.utc.edu/asee/ASEE-96-full.html.

[4] G. R. Alves, M. G. Gericota, J. B. Silva, J. B. Alves: *"Large and small scale networks of remote labs: a survey"*, University of Deusto, 2007. ISBN 978-84-9830-662-0.

[5] D. Schumacher, "Student Undergraduate Laboratory and Project Work," editorial to the special issue, *Eur. J. Phys.*, Vol. 28, No. 5, 2007.

[6] "IT Innovative Practices in Secondary Schools-Remote Experiments" Eds. Olga Dziabenko and Javier Garcia-Zubia, University of Deusto, Bilbao, Spain, 2011, ISBN: 978-84-15759-16-4.

[7] K. Azad, M. Auer, J. Harward, *Internet Accessible Remote Laboratories: Scalable E-Learning Tools for Engineering and Science Disciplines.* IGI Global (USA), 2011, ISBN 978-1-61350-186-3.

[8] F. Schauer, M. Ozvoldová and L. Tkac, INTe-L: wide open door for education by remote and virtual experiments exemplified on electricity, magnetism and electromagnetism, in *IT Innovative Practices in Secondary Schools Remote Experiments*, Chapter 9, pp. 205 –255 . eds. O. Dziabenko and J. Garcia-Zubia, University of Deusto Bilbao, Spain, 2013, ISBN: 978-84-15759-16-4.

[9] A. Rosa, "The challenge of instructional laboratories in distance education," *ABET Annual Meeting*, Baltimore, MD: ABET, 2003.

[10] K. Zakova, "Two ways of inverted pendulum remote control", *6th WSEAS International Conference on Education and Educational Technology,* Venice, Italy, eds. R. Revetria, A. Cecchi, V. Mladenov, Nov. 21-23, 2007, El. Comp. Eng., pp. 139-144.

[11] P. Bistak, "Matlab and Java Based Virtual and Remote laboratories for Control Engineering", In Proc. *17th Mediterranean Conference on Control and Automation*, Thessaloniki, Jun, 24-26, 2009, Vols. 1-3, pp.1439-1444.

[12] L. Válková, and F. Schauer, "Remote interactive real experiment in electrochemistry as exemplified on the experiment Electrochemical cell ", *6th Int. Conference on Emerging e-learning Technologies and Applications*, The High Tatras, Slovakia, September 11-13, 2008.

[13] L. Válková, F. Schauer, and M. Ozvoldova, "Electrochemical cell characterization – is it start of remote experiments in chemistry education?", In Proc. *Int. Conf. REV 2009*, Bridgeport (USA), 22-25 June 2009, International Association of Online Engineering, 2009, ISBN 978-3-89958-480-6, pp. 326-331.

[14] F. Lustig, "Computer based school experiments with measurement Intelligent

School Experimental System (ISES) under Windows, In: Proceedings *MEDACTA 97 – education in changing world*, Institute of didactic technology, PF UKF in Nitra, 1997, pp. 404-408, ISBN 80-967339-9-0

and F. Lustig, and F. Schauer, "Creative laboratory experiments for basic physics using computer data collection and evaluation exemplified on the ISES", in Proceedings of *The first European Conference on Physics Teaching in Engineering Education*, ed. Oehlenschlaeger, Copenhagen, Denmark, 1997, pp. 125-131.

[15] P. Spilakova, R. Jasek, F. Schauer, "Security risk of Java applets in remote experimentation and available alternatives", 2014International Conference on applied mathematics, computational science and engineering (AMCSE), pp. 201-212. Varna, Bulgaria, ISBN: 978-1-61804-246-0.

[16] C. Wieman, "A new model for post-secondary education", *The Optimized University*,2006 http://www.cwsei.ubc.ca/about/BCCampus2020_Wieman_think_piece. pdf .

[17] J. Bransford, A. Brown, and R. Cocking, eds., "How people learn: Brain, mind, experience, and school" *National Academy Press*, Washington DC, 2002.

[18] P. Ross, "The expert mind", *Scientific American*, Aug. 2006, and K. A. Ericsson, et al, The Cambridge Handbook of Expertise and Expert Performance, Cambridge Univ. Press, 2006, p. 64.

[19] W. K. Adams, K. K. Perkins, N. Podolefsky, M. Dubson, N. D. Finkelstein and C. E. Wieman, "A new instrument for measuring student beliefs about physics and learning physics: the Colorado Learning Attitudes about Science Survey", *Physical Review Special Topics: Phys. Educ.* Res. 2, 010101, 2006

[20] E. Redish, "Teaching Physics with the Physics Suite", Wiley 2003.

[21] S. V. Chasteen, K. K. Perkins, P. D. Beale, S. J. Pollock, and C. E. Wieman, "A Thoughtful Approach to Instruction: Course Transformation for the Rest of Us", *Journal of College Science Teaching*, Vol. 40, No. 4, 2011.

[22] J. Handelsman, D. Ebert-May, R. Beichner, P. Bruns, A. Chang, R. DeHaan, J. Gentile, et al. 2004, "Scientific teaching", *Science* 304 (5670), pp. 521-2.

[23] K. Cummings, J. Marx, R. Thornton and D. Kuhl. "Evaluating innovation in studio physics", *American Journal of Physics* 67 (7),1999. pp. 38–44.

[24] R. R. Hake, "Interactive-engagement versus traditional methods", *American Journal of Physics* 66 (1) 1998, pp. 64–74.

[25] L.C. McDermott, and E. R. Redish, "Resource letter PER-1: Physics education research", *American Journal of Physics*, 67 (9), 1999, pp.755–767.

[26] "Indicators of science and mathematics education", *National Science Foundation (NSF)*, Arlington, VA, 1996.

[27] "The Science Education Initiative of the University of Colorado" http://colorado. edu/sei.

[28] C. E. Wieman, Science Education Initiative at the University of British Columbia" http://www.cwsei.ubc.ca

[29] C. E. Wieman, "Why not try a scientific approach to Science education? " *Change: The Magazine of Higher Learning, Change,* No 5, 2007, p. 9.

[30] C. Wieman, K. Perkins and S. Gilbert, "Transforming Science Education at Large Research Universities: A case Study in Progress", *Change* 42, No2, 2010, pp. 6-14.

[31] D. Hestenes, M. Wells, G. Swackhammer, *The Physics Teacher.* 30, 1992, p.141.

[32] R. Hake, "A six-thousands-student survey", *AJP* 66, 1988, pp. 64-74.

[33] K. K. Perkins, M. M. Gratny, W. K. Adams, "Towards characterizing the relationship between students' interest in and their beliefs about physics", eds. Heron, P; McCullough, L; Marx, J, *Physics Education Research Conference* Salt Lake City, UT Date: August 10-11, 2005.

[34] Jane E. Caldwell, "Clickers in the Large Classroom: Current Research and Best-Practice Tips", CBE Life SciEduc Vol. 6 no. 1, 2007, pp. 9-20, http://www.lifescied.org/.

[35] C. E. Wieman, Wendy K. Adams, Katherine K. Perkins,"PhET: Simulations That Enhance Learning", *Science*, Vol. 322, 31 October, 2008, p. 683.

[36] C. Wieman, "Applying New Research to Improve Science Education", *Issue in science and technology*, No 1, 2012, pp. 1-7.

[37] G. Colvin, "Talent Is Overrated – What Really Separates World-Class Performers from Everybody Else", *Library of Congress cataloging-in-publication data*, Penguin Books Ltd, 2008.

[38] To be published, (System IPAL, as a module of MOODLE, was kindly provided by dr. Bill Junkin, Director of Instructional Technology at Eckerd College Tampa, St. Petersburg, Florida, USA)

[39] C. H. Crouch, J. Watkins, A. P. Fagen and E. Mazur, "Peer Instruction: Engaging Students One-on-One, All At Once", *Research-Based Reform of University Physics 2007,* http://www.compadre.org/Repository/document/ServeFile.cfm?ID=4990&DocID=241.

[40] www.phet.colorado.edu

[41] http://www.compadre.org/osp/

[42] See the webpage of Fu-K. Hwang, http://www.phy.ntnu.edu.tw/ntnujava/

[43] http://www.walter-fendt.de/ph14sk/

[44] F. Schauer, M. Ožvoldová and L. Tkáč, "INTe-L – A Wide Open Door for Education by Remote and Virtual Experiments Exemplified on Electricity, Magnetism and Electromagnetism", *IT Innovative Practices in Secondary Schools Remote Experiments*, Chapter 9, pp. 205 –255, eds. Olga Dziabenko and Javier García-Zubía, University of Deusto Bilbao 2013.

[45] T. Abdel-Salam, P. Kauffman, and G. Crossman, "Does the lack of hands-on experience in a remotely delivered laboratory course affect student learning?", *European Journal of Engineering Education*, 31,(6), 2006, pp.747–756.

[46] E. Lindsay, S. Naidu, and M. Good, "A different kind of difference: theoretical implications of using technology to overcome separation in remote laboratories" *International Journal of Engineering Education*, 23,(4), 2007, pp. 772–779.

[47] D. Lowe, S. Conlon, S. Murray, L. Weber, M. D. L. Villefromoy, E. Lindsay, "LabShare: towards cross-institutional laboratory sharing". A. Azad, M. Auer, & J. Harward Eds., *Internet accessible remote laboratories: scalable e-learning tools for engineering and science disciplines,* IGI Global 2012, pp.453–467.

[48] "Using Remote Labs in Education", Ed. J. Garcia-Zubia, and G. Alves, University of Deusto, Bilbao, Spain, 2011, ISBN: 978-84-9830-335-3,http://www.weblab.deusto.es/web/weblab.content/using_remote_labs_in_education.pdf.

[49] J. Garcia-Zubia, P. Orduna, D. Lopez-de-Ipina, G. Alves, "Addressing Softvér Impact in the Design of Remote Laboratories", *IEEE Transactions on: Industrial Electronics*, Vol. 56, Issue 12, pp. 4757-4767, 2009.

[50] F. Schauer, I. Kuřitka and F. Lustig, "Creative Laboratory Experiments for Basic Physics Using Computer Data Collection and Evaluation Exemplified on the Intelligent School Experimental System (ISES)", *Innovations 2006, World Innovations in Engineering Education and Research*, iNEER Special Volume, W. Aung et al. (eds.), pp. 305-312, USA, 2006.

[51] F. Schauer, M. Krbecek and M. Ožvoldová, "Controlling Programs for Remote Experiments by Easy Remote – ISES (ER-ISES)", *Proceedings of the International Conference on Remote Engineering and Virtual Instrumentation – REV 2013*, Sydney, Australia, 6-8 February 2013.

[52] F. Schauer, F. Lustig and M. Ozvoldova, "ISES – Internet School Experimental System for Computer-Based Laboratories in Physics", *Innovations 2009, World Innovations in Engineering Education and Research*, iNEER Special Volume 2009, (USA), W. Aung et al. (eds.), pp.109-118.

[53] Wikipedia, "Finite-State Machine," http://en.wikipedia.org/wiki/Finite-state_machine, accessed on 24.11.2014.

[54] http://cordis.europa.eu/fp7/ict/programme/challenge8_en.html (accesses on 3. 1. 2014).

[55] http://en.wikipedia.org/wiki/Plug_and_play, (accessed on 1.7.2014).

[56] D. Lowe, P. Newcombe and B. Stumpers, "Evaluation of the Use of Remote Laboratories for Secondary School Science Education", Res Sci. Educ. 43 (2013), 1197–1219.

[57] J. Ma, and J. V. Nickerson, "Hands-On, Simulated, and Remote Laboratories: A Comparative Literature Review," *ACM Computing Surveys*, Vol. 38,2006.

[58] J. Corter, J. V. Nickerson, S. K. Esche, C. Chassapis, ImSeongah, and J. Ma, "Constructing Reality: A Study of Remote, Hands-On, and Simulated Laboratories," *ACM Transactions on Computer-Human Interaction*, Vol. 14, No. 2, Article 7, 2007.

[59] T. Restivo, F. G. Almeida, M. F. Chouzal, J. G. Mendes, A. M. Lopes, J. P. Carneiro, and H. S.Lopes, "On the Way for a Better Methodology in Teaching/Learning Instrumentation for Measurements at Mechanical Engineering Compulsory Syllabus", in *Proc. WCETE*, 2004, pp. 92–96.

[60] Z. Doulgeri, T. Matiakis, "A Web Telerobotic System to Teach Industrial Robot pPath Planning and Control," *IEEE Trans. Educ.*, Vol. 49, No. 2, pp. 263–270, 2006.

[61] G. Canfora, P. Daponte, and S. Rapuano, "Remotely Accessible Laboratory for Electronic Measurement Teaching," *Comput. Standards and Interfaces* 26, 6, pp. 489–499, 2004.

[62] F. Lustig F. Schauer, M. Ožvoldová, "Plug and play system for hands-on and remote laboratories". In: *Proceedings of 8th International Conference on Hands-on Science*. Ljubljana: University of Ljubljana, 2011, pp. 17-21 ISBN 978-989-95095-7-3.

[63] M. Bochicchio, A. Longo, "Extending LMS with Collaborative Remote Lab Features," *ICALT*, 2010, pp.310-314.

[64] D. Gillet, E. Law, and A. Chatterjee, "Personal Learning Environments in a Global Higher Engineering Education," Web 2.0 Realm, *1st IEEE Engineering Education Conference (EDUCON)*, Madrid, Spain, April 14-16, 2010.

[65] E. G. Guimarães, E. Cardozo, D. H. Moraes, and P.R. Coelho, "Design and Implementation Issues for Modern Remote Laboratories," *IEEE Transactions on Learning Technologies*, Vol. 4, No. 2, 2011, pp. 149-161.

[66] F. Schauer, M. Ožvoldová and F. Lustig, "Integrated e-Learning – New Strategy of Cognition of Real World, in Teaching Physics", *Innovation 2009, World Innovations in Engineering Education and Research*, iNEER Spec. Vol., W. Aung et al. (eds.), 2009, USA, pp. 119-136.

[67] W. Christian, F. Esquembre, Phys.Teach., Vol. 45, November 2007, 475.

[68] F. Schauer, F. Lustig and M. Ozvoldova, *"Internet Natural Science Remote e-Laboratory (INRe-L) for Remote Experiments"*: in Teaching Physics, *Innovation 2009, World Innovations in Engineering Education and Research*, iNEER Spec. Vol., W. Aung et al. (eds.), 2011, USA, pp. 51-68.

[69] C. Wieman, K. Perkin,"Transforming Physics Education", *Physics Today*, Vol. 58, Nov. 2005, pp. 26-1.

[70] F. Schauer, F. Lustig, J. Dvořák and M. Ožvoldová, "An easy-to-build remote laboratory with Data Transfer Using the Internet School Experimental System", *Eur. J. Phys*, Vol. 29, 2008, pp. 753-765.

[71] S. Gröber M. Vetter, B. Eckert and H. J. Jodl, "Experimenting from a Distance – Remotel Controlled Laboratory (RCL)", *Eur. J. Phys.* Vol. 28, No. 5, 2007, p. 127.

[72] C. Wieman and K. Perkins, "A powerful tool for teaching science", *Nature physics*, Vol. 2, 2006, p. 290.

[73] R. Mayer, "Learning and Instruction", Merrill, Upper Saddle River, NJ 2003.

[74] F. Lustig, F. Schauer, M. Ozvoldova, *E-Labs in Engineering Education: Classical, Real Remote or Virtual? In PROCEEDINGS of the Conference ICTE 2007"* Publ. University of Ostrava, 2007,pp. 107-116. 17. 9. 2007, Rožnov pod Radhoštěm, CZ, ISBN 978-80-7368-388-7.

[75] N. D. Finkelstein, W. Adams, C. Keller, K. Perkins, C. Wiemanand the PhET Team, "High-Tech Tools for Teaching Physics: the Physics Education Technology Project, *Journal of Online Learning and Teaching, iJOE* Vol. 2, No. 3, September 2006, p. 109.

[76] C. Thomsen, S. Jeschke, O. Pfeiffer and R. Seiler, *e-Volution: eLTR – Technologies and Their Impact on Traditional Universities, Proceedings of the Conference: EDUCA online*, ISWE GmBH, Berlin 2005.

[77] F. Schauer, M. Ozvoldova, F. Lustig, Real Remote Physics Experiments across Internet – Inherent Part of Integrated E-Learning, *iJOE*, Vol 4, No 2 (2008), 52-55, ISSN 1861-2121.

[78] M. Ožvoldová and F. Schauer: Remote Experiments in Freshman Engineering Education by Integrated e-Learning, Internet Accessible Remote Laboratories:

Scalable E-Learning Tools for Engineering and Science Disciplines, ed. Abul K.M. Azad, Michael E. Auer and V. Judson Harward, IGI Global 2011, ISBN13: 9781613501863, ISBN10: 1613501862.

[79] M. Dougmias, "The use of Open Source software to support a social constructionist epistemology of teaching and learning within Internet-based communities of reflective inquiry PhD", Thesis, Science and Mathematics Education Centre Curtin University of Technology Perth, Western Australia https://dougiamas. com/archives/

and M. Dougmias, P. Taylor, "Moodle: Using Learning Communities to Create an Open Source Course Management System." *The 15th World Conference on Educational Multimedia and Hypermedia and World Conference on Educational Telecommunications*, 2003, edited by AACE, 171-78. Norfolk, VA: ED-MEDIA 2003.

[80] http://moodle.org

[81] W. Christian, M. Belloni, "Physlet® Physics: Interactive Illustrations, Explorations and Problems for Introductory Physics", Davidson College ISBN-10: 0131019694, http://webphysics.davidson.edu/applets/Applets.html.

[82] M. Ožvoldová, P. Čerňanský, F. Schauer, F. Lustig, "Internet Remote Physics Experiments in Student's Laboratory", *INEER, Innovation 2006, World Innovations in Engineering Education and Research,* iNEER Special Volume,Virginia, USA – Chapter 25, pp. 297-304, ISBN 0-9741252-5-3.

[83] M. Ožvoldová, M. Žovínová, Remote Experiments – the Latest Technologies in Physics Classes, Journal of Technology and Information education, Vol. 3, No. 1 (2011), pp. 26-32, ISSN 1803-537X.

[84] M. Ožvoldová, F. Schauer, "Remote Experiments in Freshman Engineering Education by Integrated e-Learning", *Internet Accessible Remote Laboratories: Scalable E-Learning Tools for Engineering and Science Disciplines* ", pp. 60-83 pp, IGI Global, USA, 2011, ISBN 978-1-61350-187-0.

[85] M. Žovínová, M. Ožvoldová, 2011. Remote temperature measurement – project-based learning. In Exp.at'11, Lisbon, Portugal: Calouste Gulbenkian, ISBN 978-989-20-3008-1. pp. 62-66.

[86] M. Ožvoldová, M. Žovínová 2011. Remote Experiments – the Latest Technologies in Physics Classes. In Journal of Technology and Information Education [online]. 2011, Vol. 3, No. 1. (2011), p. 26-32 Available on: http://www.jtie.upol.cz/clanky_1_2011/ozvoldova.pdf, ISSN 1803-537X.

[87] M. Kostelníková, Possibilities of Remote Experiments Utilisation in Science Education, Thesis (In Slovak "Možnosti využitia vzdialených experimentov v prírodovednom vzdelávaní"), Trnava University in Trnava, 2013, 160 pp.

[88] F. Schauer, P. Majerčík, Real Interactive Pendulum Experiment with Data Collection and Transfer across Internet, C. Cassan, (Ed.): MPTL 14, Proceeding of International Conference Multimedia in Physics Teaching and Learning Conference, MPTL 2009, Litho Stampa, University of Udine, Italy, 2009, pp. 74-75, ISBN 2-914771-61-4.

[89] F. Schauer, M. Ožvoldová, P. Čerňanský, T. Kozik, L. Válková, A. Slaninka, M. Kostelníková, P. Majerčík, L. Tkáč, Slovak e-Laboratory of Remote Interac-

tive Experiments for University Teaching by Integrated e-Learning strategy / In Proc. of 6th International Conference on Emerging eLearning Technologies and Applications, ICETA 2008, September 11 – 13, 2008, Stará Lesná, The High Tatras, Slovakia, pp. 467-472, ISBN 978-80-8086-089-9.

[90] M. Ožvoldová, I. Červeň, P. Čerňanský, J. Dillinger, S. Halúsková, O. Holá, P. Fedorko, I. Štubňa J. Krajčovič, P. Ballo, I. Jančuška, J. Krempaský, K. Kvetaň, M. Beňo, Multimedia University Physics Textbook 2, (In Slovak: Multimediálna vysokoškolská učebnica fyziky, časť 2), Trnava 2007, Trnava University, CD, ISBN 978-80-8082-128-9.

[91] P. Brom and F. Lustig: Use of Integrated e-Learning for quantum mechanics introduction (in Czech Využití metody integrovaného e-Learningu pro úvod do kvantové fyziky). In Ulrch, Zatloukal: 9. International conference Alternative methods of teaching 2011, Hradec Králové University, 1011, ISBN 978-80-104-4.

[92] V. J. Harward et al., „The iLab Shared Architecture: A Web Services Infrastructure to Build Communities of Internet Accessible Laboratories," *Proceedings of the IEEE,* Vol. 96, 2008, pp. 931-950.

[93] http://www.lila-project.org/, accessed on 1. 8. 2014.

[94] D. Lowe, S. Murray, E. Lindsay, and L. Dikai, "Evolving Remote Laboratory Architectures to Leverage Emerging Internet Technologies," *Learning Technol., IEEE Trans.* Vol. 2, 2009, pp. 289-294.

[95] M. Tawfik, D. Lowe, S. Murray, M. de la Villefromoy, M. Diponio, E. Sancristobal, M. José Albert, G. Diaz, M.Castro, "Grid Remote Laboratory Management System – Sahara Reaches Europe," *Proc. 10th REV conference*, Sydney, 2013.

[96] http://www.go-lab-project.eu/, accessed on 1. 8. 2014.

[97] F. Schauer, M. Krbecek, P. Beno, M. Gerza, L. Palka and P. Spilaková, "REMLABNET – open remote laboratory management system for e-experiments", *REV 2014*, Porto, 26-28 February 2014.

[98] D. Lowe, P. Newcombe and B. Stumpers, "Evaluation of the Use of Remote Laboratories for Secondary," *Res. Sci. Educ/Science Education*, Springer Science and Business Media B.V. 2012, DOI 10.1007/s11165-012-9304-3.

[99] M. Krbecek, F. Schauer and F. Lustig, "EASY REMOTE ISES – Environment for Remote Experiments Programming," *Innovations 2013: World Innovations in Engineering Education and Research*, W. Aung, et al. (eds.), iNEER, Potomac, MD, USA, pp. 80 – 101.

[100] T. Dulík, M. Bližňák, "Security measures in virtual laboratory of microprocessor technology," DAAAM International *Vienna, Proceedings of the 21st International DAAAM Symposium "Intelligent Manufacturing & Automation: Focus on Interdisciplinary Solutions"*, Vienna, 2010, pp. 1203-1204, ISBN-ISSN 978-3-901509-73-5.

[101] J. Vojtěšek, M. Bližňák, R. Matušů, T. Dulik, "Virtualization as a Teaching Tool for IT-based Courses," *WSEAS World Science and Engineering Academy and Science, WSEAS Transactions on Advances in Engineering Education*, Athens, 2009, pp. 265-274, ISSN 1790-1979.

Assoc. Prof. Dr. Miroslava Ožvoldová received her M.S. degree in Physics from Comenius University in Bratislava, Slovakia, in 1973, and in 1981 a Ph.D. in Physics-Mathematics Science. In 1992 she was appointed as Associate Professor at the Faculty of Materials Science and Technology in Trnava and Slovak University of Technology in Bratislava. Since 2003 she has been active at the University of Trnava, Faculty of Education. Since 2008 she is also with the Faculty of Informatics, Tomas Bata University in Zlin, Czech Republic. Her main activities are optical properties of chalcogenide and heavy metal optical glasses and the utilisation of a new generation of e-learning in education processes with the support of e-experiments (real remote experiments and interactive simulations) in Physics teaching. She is the author of about 190 papers in both Solid state physics and in the field of Physics education of pre-service and in-service physics teachers.

Prof. Dr. František Schauer received a M.S. degree in Electronics from the Brno University of Technology in 1963 and his Ph.D. degree in Solid State Physics from Prague University of Technology in 1978. In 1982 he was appointed as Associate Professor and in 1988 Professor in Condensed Matter Physics at the Technical Academy in Brno, Czech Republic. In 1993-2002 he was with the Faculty of Chemistry, Brno University of Technology and then he was with the Polymer Centre of the Faculty of Technology. At present he is with the Faculty of Informatics, Tomas Bata University in Zlin, Czech Republic and Faculty of Education, Trnava University of Trnava, Slovak Republic. His main activities are molecular organic electronics, electronic structure spectroscopy by electrochemical and charge injection methods and computer assisted experiments. He is the author of about 200 papers in both Solid state physics and Education practice with about 350 SCI citations.